Exploring Science

Program Consultants

Randy L. Bell, Ph.D.

Malcolm B. Butler, Ph.D.

Kathy Cabe Trundle, Ph.D.

Judith S. Lederman, Ph.D.

Center for the Advancement of Science in Space, Inc.

Welcome to Exploring Science 2

Keeping a Science Notebook 4

Set up Your Science Notebook 6

Why Explore Science? 7

Nature of Science ... 8

What Is Science? ... 10

How Do Scientists Work? ... 12

Who Are Scientists? ... 14

Investigate **Practice Science** ... 16

Let's Explore! ... 18

Physical Science 20

Energy

Batter Up! 22

Investigate Speed 24

Hit the Ball 26

Investigate Collisions 28

Sounds of the Game 30

Investigate Sound 32

The Sun's Light 34

Investigate Light 36

Stories in Science Captured by Light 38

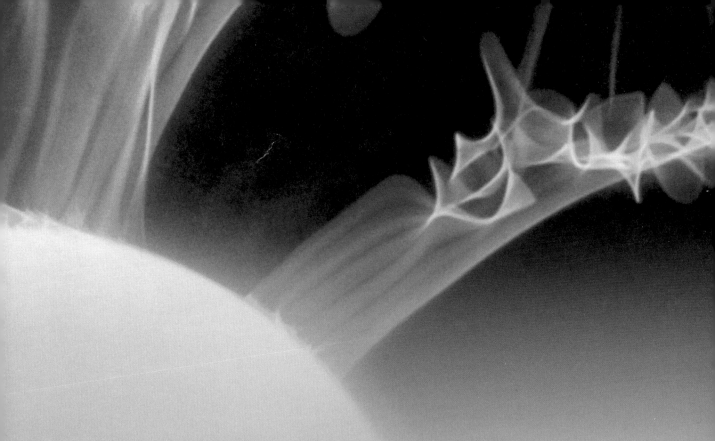

Heat It Up! 40

Investigate Heat 42

It's Electric 44

Electric Circuits 46

Investigate Electric Circuits 48

Spin It! 50

STEM Space Station Project 52

Think Like an Engineer *Case Study* Finding Solutions to Energy Problems 56

Think Like an Engineer Design, Test, and Refine a Device 62

Think Like an Engineer Design, Test, and Refine a Device 66

Nonrenewable Energy Resources 68

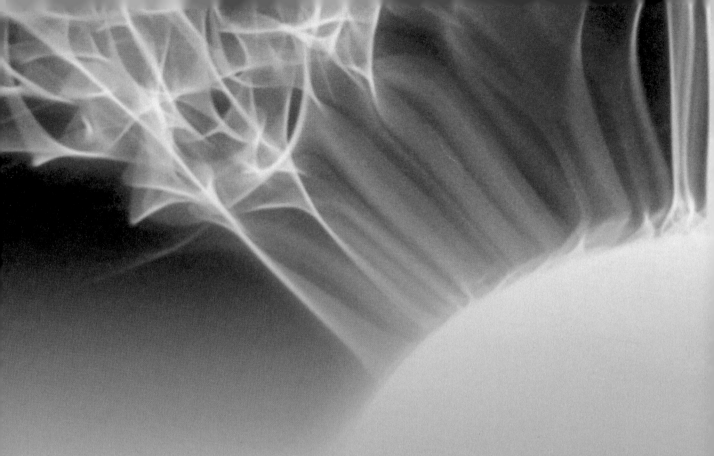

Physical Science (continued)

Renewable Energy Resources 70

Energy Resources and the Environment 72

Think Like a Scientist Obtain and Combine Information 74

Waves: Waves and Information

Waves 76

Properties of Water Waves 78

Properties of Sound Waves SCIENCE in a SNAP 80

Investigate How Waves Move Objects 82

Investigate Wavelength and Amplitude 84

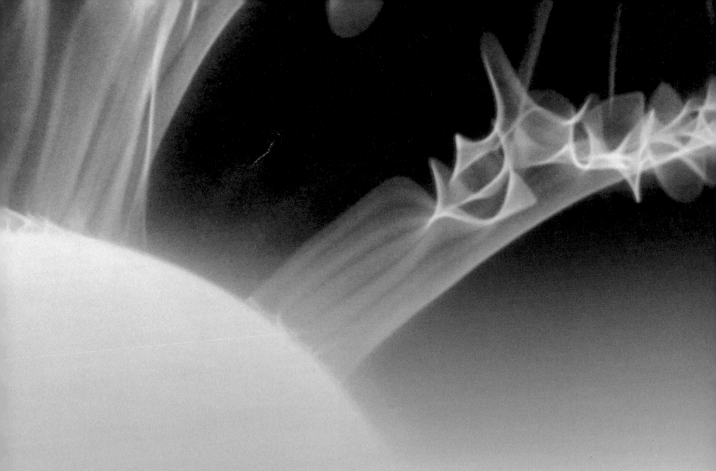

Information Technology—GPS	86
Information Technology—Cell Phones	88
Investigate Use a Code	90
Think Like an Engineer Compare Multiple Solutions	92
Stories in Science Code Talkers	94
STEM Engineering Project	96
Science Career Animal Tracker	100
Citizen Science Track Bird Life	102
Check In	104
Let's Explore!	106

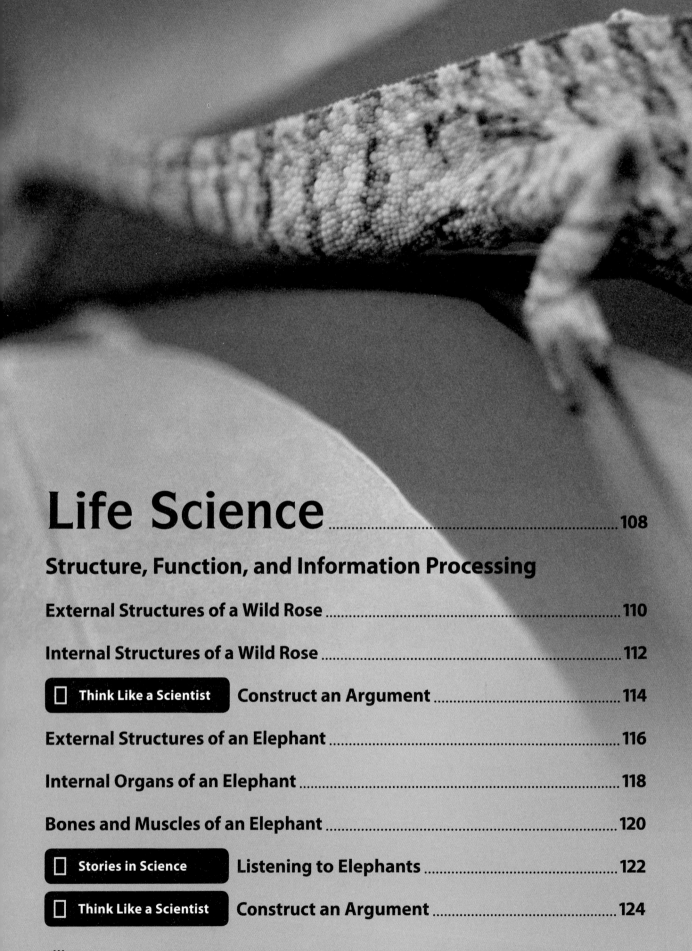

Life Science ... 108

Structure, Function, and Information Processing

External Structures of a Wild Rose ... 110

Internal Structures of a Wild Rose ... 112

Think Like a Scientist Construct an Argument ... 114

External Structures of an Elephant ... 116

Internal Organs of an Elephant ... 118

Bones and Muscles of an Elephant ... 120

Stories in Science Listening to Elephants ... 122

Think Like a Scientist Construct an Argument ... 124

Animal Senses 126

Light and Sight 128

　Investigate　**How We See** 130

　Think Like a Scientist　**Use a Model** 132

　STEM　**Research Project** 134

　Science Career　**Dog Whisperer** 138

Check In 142

　Let's Explore! 144

Earth Science 146

Earth's Systems: Processes that Shape the Earth

Rainfall in the United States 148

Pacific Northwest Forest 150

Southwest Desert 152

Central Plains Grassland 154

Eastern Temperate Forest 156

Weathering 158

Erosion and Deposition 160

Wind Changes the Land 162

Water Changes the Land 164

Investigate Weathering and Erosion 166

Ice Changes the Land 168

Living Things Change the Land 170

Landslides Change Earth's Surface 172

Think Like an Engineer Make Observations 174

Natural Hazards 178

Earthquakes 180

Investigate Earthquakes 182

Tsunamis 184

Volcanoes 186

Reducing the Impact of Natural Hazards 188

x

Early Warning Systems 190

Tsunami Detection 192

   STEM **Engineering Project** 194

Patterns of Water and Land Features 198

   ☐ Think Like a Scientist   Analyze and Interpret Data 200

   ☐ Think Like an Engineer *Case Study*   Building for the Future 202

   ☐ Think Like an Engineer   Generate and Compare Solutions 206

The Badlands 210

Iceland 212

   ☐ Think Like a Scientist   Identify Evidence 214

   ☐ Stories in Science   A Feel for Fossils 218

   ☐ Science Career   Crisis Mapper 220

Check In 222

   ☐ More to Explore 223

Science Safety 224

Tables and Graphs 226

Glossary 230

Index 236

Credits, Consultants, and Copyright 244

NATIONAL GEOGRAPHIC | Explorer

David Moinina Sengeh is a biomedical engineer, a data scientist, and a National Geographic Explorer. David designs equipment for people with injuries. Wearing the equipment he designs helps them move more easily.

Welcome to Exploring Science!

Hello, explorers! I am David Moinina Sengeh. Welcome to *Exploring Science*! Together we are going to learn how science helps us solve problems and make new discoveries.

A big part of my own work as a scientist and engineer involves designing artificial limbs. A person who has lost a leg can be fitted with an artificial one that moves very much like a human leg.

As a scientist, I love solving problems. I grew up in Sierra Leone, West Africa, during a civil war. I saw many people with injuries as a result of the fighting. Some of these people were given artificial limbs to wear. But I noticed many chose not to. That made me wonder, why not? Through my investigation, I learned that the artificial limbs were uncomfortable and even painful to wear. They were poorly designed and did not fit well.

I made it my goal to design artificial limbs that are fitted to each individual and comfortable to wear. I want my solutions to help people get the most out of life.

I use technology, test new ideas, and work with engineers to improve my designs. In *Exploring Science*, you will practice many of the same skills I use. You will solve problems and make discoveries, too!

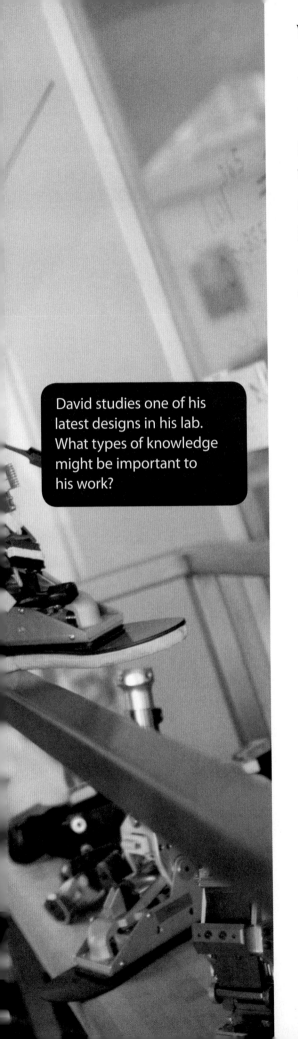

David studies one of his latest designs in his lab. What types of knowledge might be important to his work?

Keeping a Science Notebook

It is important for me to have a record of my work. I keep a science notebook on my computer. I type up my notes, measurements, and results. I add questions and ideas that come to me. I can go back in my notebook to study this information. I can look for patterns and make predictions. I can form explanations.

You can keep your own science notebook. What will you record? This list and the examples on the next page can give you ideas.

- Define and draw science words and main ideas.
- Label drawings. Include captions and notes to explain ideas.
- Collect objects, such as photos and magazine or newspaper clippings.
- Add tables, charts, or graphs to record observations and data.
- Record evidence for explanations and conclusions.
- Think about what you've done and learned. Ask new questions.

Now you are ready to set up your science notebook! What ideas of your own might you have?

David takes time to update his science notebook on his computer.

▼ Use drawings to help you remember what you learn.

▼ Draw models to explain main ideas.

Energy from Wind

A Wind Turbine
Energy from wind turns the wind turbine. Energy of motion is changed to electrical energy. This makes less pollution than burning fossil fuels.

Parts of an Electric Circuit

wire
battery
bulb
switch

Main idea: In an electric circuit, wires connect a battery to a switch and a light bulb. Turning on the switch completes the path for electric current to flow from the battery to the light bulb.

▶ Record new questions and look for answers.

My Earth Science Questions

1. How much rain falls each year in the area where I live?

2. Do deciduous trees stop making food during winter?

3. Where are glaciers still found on Earth today?

4. How much warning do people usually get before an earthquake happens?

Set up Your Science Notebook 📓 My Science Notebook

Use your science notebook every time you study science. Here are a few suggestions to make your notebook unique and easy to use. Your teacher may have more instructions. Use your own ideas, too!

- Design a cover. Include something you like about science or something you would like to learn. Be sure to add your name.
- In the front of your notebook, write "Table of Contents." Leave some blank pages. For each entry, you will need to write the date, title, and a page number.
- Organize your notebook. Add tables and graphs. Label everything carefully. Write the date on every new entry and number each new page.
- Keep your science notebook in a safe place. You'll want it to last, so you can see and share how much you have learned.

▽ Design a cover that is all about science and you!

▽ Make a table of contents on the second page. Add information as you read and investigate.

Why Explore Science?

Science is a way of learning about nature. It is a way of discovering and explaining how things work the way they do. And when you learn more about one area of science, you can often apply it to another. Many of the laws of nature can be applied throughout the universe.

Have you ever noticed something in nature and wondered about it? Did you explore more to form an explanation? If so, you were acting like a scientist! You were using the scientific practices of observing, questioning, and investigating.

In *Nature of Science,* you will learn more about these practices. You will learn what science is and what it isn't. You will read about different ways that scientists work and one thing that all scientists rely on: evidence. As you learn about scientists at work, what questions do you have? Write them in your notebook!

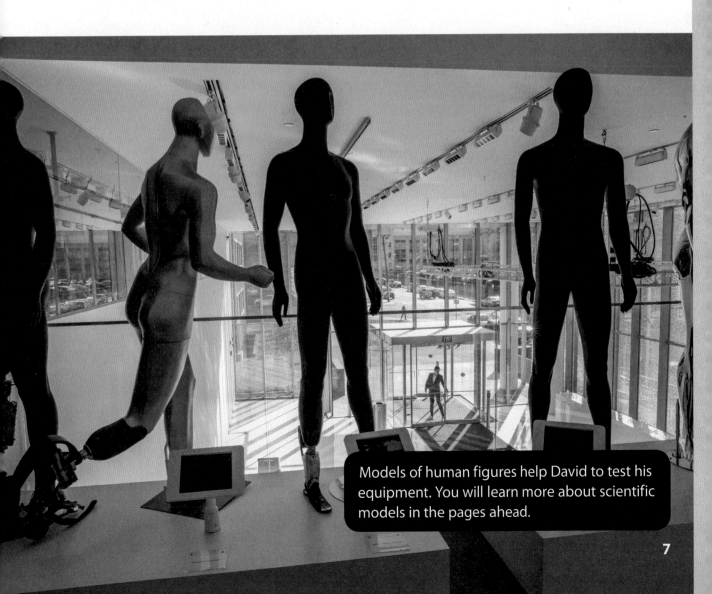

Models of human figures help David to test his equipment. You will learn more about scientific models in the pages ahead.

Nature of Science

From the air, the San Andreas Fault in California looks like a giant slash across Earth's surface. A fault line is a sign of powerful movements within Earth's crust.

What Is Science?

You might think doing science means working in a lab or studying facts in books. You are only partly right! Science is a way of knowing about the natural world. It is a process of discovery that never stops. And anyone can do it!

Scientific thinking often begins with simple observations. To **observe** means to use your senses to collect information about the world around you. Observations often lead to questions that scientists try to answer. Study the photos on this page and on the previous page. What do you observe? What questions do you have?

These photos show places where earthquakes happen. An earthquake is a natural event in which the ground shakes, sometimes violently. You may already know a bit about earthquakes. You will learn more about them in this book.

Humans have always tried to explain natural events, such as earthquakes. Ancient people did not have tools to directly study what caused the ground to shake. They often used their beliefs to explain what they observed. Some said giant creatures moving under Earth's surface caused earthquakes!

Science knowledge is not based on beliefs. Instead it is based on evidence. **Evidence** is information that comes from analyzing or finding patterns in observations and data such as numbers and measurements. Over time, scientists and engineers have developed tools to accurately study earthquakes. They have collected data and made inferences about the data to build a body of evidence. To **infer** is to draw conclusions based on new information and facts that are already known. Scientists are still collecting evidence today and forming new and improved explanations.

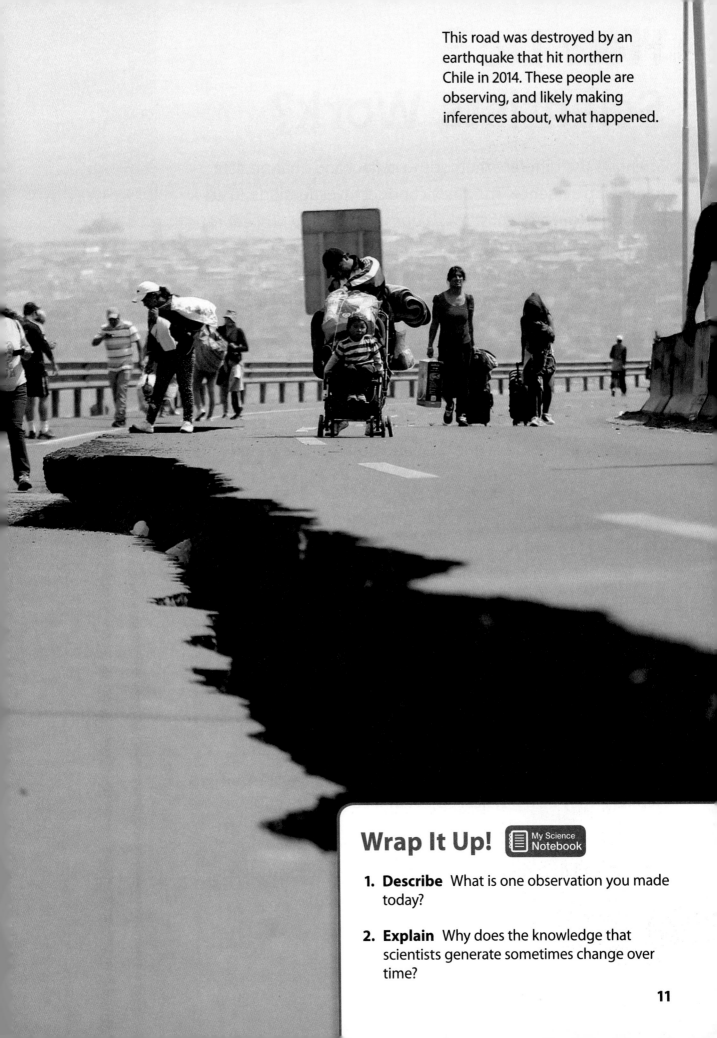

This road was destroyed by an earthquake that hit northern Chile in 2014. These people are observing, and likely making inferences about, what happened.

Wrap It Up!

1. **Describe** What is one observation you made today?

2. **Explain** Why does the knowledge that scientists generate sometimes change over time?

How Do Scientists Work?

Scientists use different methods to collect and analyze data to develop evidence. They may conduct investigations. In an **investigation,** a scientist asks a question, plans a procedure to observe and collect data, and forms a conclusion. Investigations can happen in the field, where scientists make direct observations. Investigations can happen in a lab, too.

Scientists may carry out a type of investigation called an experiment. An **experiment** is a fair test, or a process in which scientists control variables to test a hypothesis. A **variable** is a factor in an experiment that may change. A **hypothesis** is an idea or explanation that can be tested by investigation.

Seismologists may experiment to test a hypothesis about the effects of earthquakes on buildings. They may change one variable, such as the shape of the building or materials used to build it. They will analyze the data to determine whether or not the results support their hypothesis.

Experiments may involve the use of models. In science, a **model** is a representation that can be used to explain or predict a natural system. Some objects or processes are difficult to study directly. They may be very big, very small, or even dangerous. Models can help. A model may be a picture. It may be a physical object or a computer program.

It is important for scientists to repeat their tests many times to make sure they have accurate results. Their tests must also be designed so that others can reproduce them.

NS Scientific Knowledge Is Based on Empirical Evidence. Science findings are based on recognizing patterns. (4-PS4-1)
NS Scientific Knowledge Assumes an Order and Consistency in Natural Systems. Science assumes consistent patterns in natural systems. (4-ESS1-1)

Tools and technology help scientists make more accurate observations. This seismologist is using equipment that measures movements caused by earthquakes.

Wrap It Up!

1. **Identify** What is an experiment?

2. **Explain** How is using a model helpful to scientists?

3. **Apply** Why is it important that a scientist can repeat another scientist's experiment?

Who Are Scientists?

Scientists may come from different backgrounds and train for different science careers. Some scientists study living things. Others study Earth's land, air, and water. Some scientists study space. Still others study materials used to make the vehicles and structures we use every day.

Scientists from different fields may share an interest in solving a problem or answering a question. For this reason, scientists often work in teams. Each scientist brings his or her own experience and training to the investigation.

Scientists from different fields also have common ways of working. For example, all scientists make observations and ask questions. Scientists investigate to collect information and analyze it to draw conclusions. Scientists use logical thinking and rely on evidence to form explanations.

Dr. Kate Hutton is a good example of how scientists can share their experience with others. During her career as a seismologist, Dr. Hutton often used her training to help the public understand earthquake events. She became known as the "Earthquake Lady."

Scientists also tend to be curious about the world around them. They use creativity and imagination to find new ways of seeing and doing things. Scientists bring their own ideas that can influence their results and also use creativity when trying to find solutions to problems and answers to questions.

Are you ready to think like a scientist? Then let's go!

Scientists and engineers teamed up to design this model of an earthquake-resistant building. They are going to shake it to put it to the test.

Wrap It Up!

1. **Summarize** What are some things that most scientists have in common?

2. **Apply** Give an example of how science affects your life every day.

Investigate

Practice Science

? **How can you think like a scientist to study something you cannot see?**

Scientists have developed a large body of evidence about Earth's interior. They know the materials that make up each of its layers and how thick each layer is. They can even describe Earth's core and its temperature. Have scientists traveled to the center of our planet to observe it for themselves? No. But they have used equipment to extend their senses and collect information about parts of Earth they cannot see. They have used observations and evidence to make inferences about Earth's interior.

A probe is a piece of equipment that can move through water, soil, or rock and has sensors that collect information. In this investigation, you will use a toothpick in much the same way as a probe. Your toothpick will help you make observations and inferences about what's inside a "mystery container."

Materials

mystery container

toothpick

SEP Engaging in Argument from Evidence. Construct an argument with evidence, data, and/or a model. (4-LS1-1)
CCC Patterns. Patterns can be used as evidence to support an explanation. (4-ESS1-1), (4-ESS2-2)
NS Scientific Knowledge Is Based on Empirical Evidence. Science findings are based on recognizing patterns. (4-PS4-1)

My Science Notebook

1 Begin by looking at your mystery container. Record your observations in your science notebook. What can you infer about the contents of the container just by looking at it? Record your ideas.

2 You cannot open your container. But you can probe what's inside with your toothpick. What can you infer based on what you sense by using your probe? Record your ideas.

3 Shake your container. Listen for any sounds that might help you infer what is inside. What can you sense about the way the object or objects move inside the container? Record your observations.

4 What other observations can you make to help you infer what is inside your container? Be sure to use your other senses, as much as your teacher allows. Record notes about anything you try.

5 Write an explanation for what you conclude is in your container. Use evidence from your investigation to support your explanation.

6 Share your explanation with your classmates. Compare your approach and results with theirs. Which explanations do you think are most accurate and why?

What might be learned by sending a probe into this underwater fault in Iceland?

Wrap It Up! **My Science Notebook**

1. **Describe** Which senses did you use to observe your mystery container?

2. **Compare and Contrast** How does your work in this investigation compare to how scientists collect data about Earth's interior?

3. **Analyze** How did this investigation help you understand what science can and cannot do?

 **Explorer**

David Moinina Sengeh
Biomedical Engineer
National Geographic Explorer

Let's Explore!

In *Exploring Science,* you have learned that scientists often conduct experiments to test hypotheses. Scientists often carry out experiments to test the effect of a change in a variable. I run experiments to test the artificial limbs I design. One variable I may change is the material used to make part of a limb. I test which material gives the best results. As you read, look for examples of scientists doing experiments. That includes you, too!

My investigations of different materials relate to physical science. Physical science is the study of matter and energy. Here are some questions you might investigate as you read *Physical Science*:

- How can a pitcher control the speed of a baseball?
- How can sunlight change a piece of construction paper?
- How is energy transferred from the sun to Earth?
- Why are coal and oil called "fossil fuels"?
- How does GPS work? How is it used?

Look at the notebook examples for ideas of other questions to ask. Let's check in again later to review what you have learned!

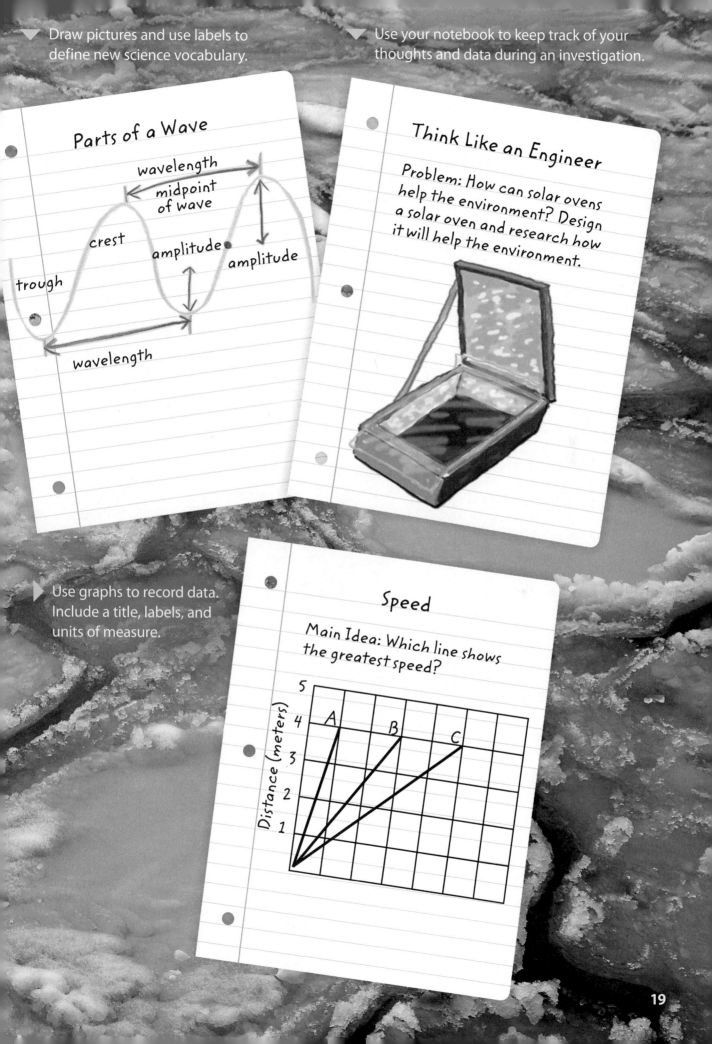

Physical Science

Energy

Waves: Waves and Information

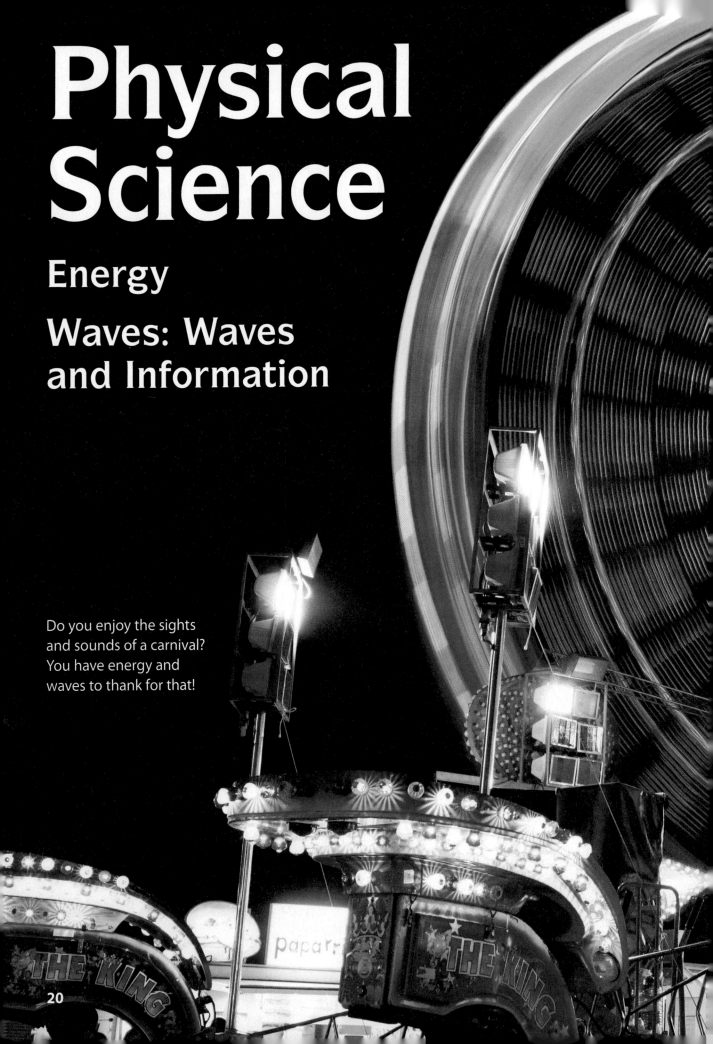

Do you enjoy the sights and sounds of a carnival? You have energy and waves to thank for that!

Batter Up!

A baseball hurls toward you at a tremendous speed. You swing... and miss! The ball was moving faster than you thought!

When a ball flies through the air, **energy** is moved from one place to another. All moving objects have energy. The faster they are moving, the more energy they have.

The pitcher can affect the ball's energy of motion by how hard he throws.

DCI PS3.A: Definitions of Energy. The faster a given object is moving, the more energy it possesses. (4-PS3-1) • Energy can be moved from place to place by moving objects or through sound, light, or electric currents. (4-PS3-2), (4-PS3-3)
CCC Energy and Matter. Energy can be transferred in various ways and between objects. (4-PS3-1), (4-PS3-2)

Pitchers can decide how fast they want to throw a baseball. A pitcher can give the ball more energy to increase its speed, or less energy to decrease its speed. If a hitter thinks the ball is moving faster or slower than it is actually moving, the hitter will miss. And that is just what the pitcher wants!

Pitchers can adjust their grip on the ball to throw it at different speeds. A typical fastball can move more than 145 km/h (90 mph).

A typical changeup moves slower, up to 145 km/h (90 mph).

Wrap It Up!

1. **Explain** How can energy be moved from place to place?

2. **Compare** How does the energy of a fastball compare with the energy of a changeup? How do you know?

Investigate

Speed

? **How is the speed of an object related to its energy?**

The fastest baseballs thrown clock unbelievable speeds over 161 km/h (100 mph). Imagine the energy such a ball possesses! Pitchers who throw the fastest pitches can even sustain injuries from using so much force. In this investigation, you can compare the speeds of balls rolled with different amounts of force.

Materials

- whiffle ball
- stopwatch

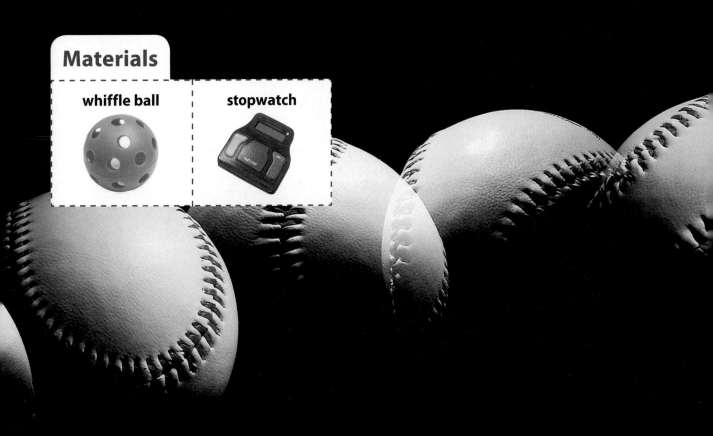

My Science Notebook

1 With a partner, kneel about 3 meters from a wall or other surface. Place a ball on the starting line. Have your partner ready to use the stopwatch.

2 Say "Go" and gently release the ball so it moves very slowly. Have your partner time how long it takes for the ball to roll to the wall. Record the time.

3 Repeat steps 1 and 2, this time rolling the ball with slightly more force.

4 Repeat steps 1 and 2, this time rolling the ball with even more force.

Wrap It Up!

1. **Describe** The less time it takes the ball to roll to the wall, the greater its speed. Describe the speeds of the ball in your three trials.

2. **Explain** Use your results as evidence to explain how the speed of an object is related to its energy.

Hit the Ball

A fastball speeds toward a batter. He hits it! The ball is in **motion**—changing position. As the ball collides with the bat, energy is **transferred** from the bat to the ball. As a result, the ball's direction changes, and it soars toward the outfield. The hitter races to first base.

The batter's swing affects the bat's energy when it collides with the ball.

DCI PS3.B: Conservation of Energy and Energy Transfer. Energy is present whenever there are moving objects, sound, light, or heat. When objects collide, energy can be transferred from one object to another, thereby changing their motion. In such collisions, some energy is typically also transferred to the surrounding air; as a result, the air gets heated and sound is produced. (4-PS3-2), (4-PS3-3)
DCI PS3.C: Relationship Between Energy and Forces. When objects collide, the contact forces transfer energy so as to change the objects' motions. (4-PS3-3)
CCC Energy and Matter. Energy can be transferred in various ways and between objects. (4-PS3-2), (4-PS3-3)

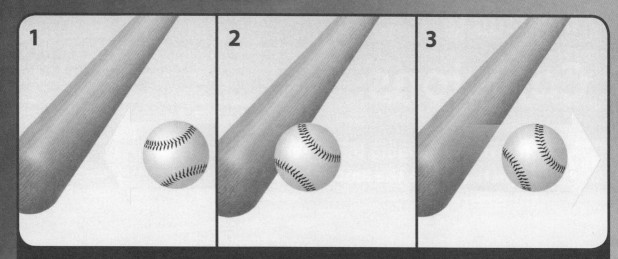

1) The ball has energy as it moves toward the bat. The bat also has energy as it moves toward the ball. **2)** When the bat and ball collide, energy is transferred between the bat and the ball at the point of contact. **3)** The motion of the ball and bat change. The bat slows down, and the ball moves in a new direction and may be moving at a new speed, too.

When objects collide, they can change each other's motion in different ways. For example, the direction of a ball's motion changes when the ball bounces off a wall. The speed of a ball's motion changes when the ball encounters air and wind and slows down. And the ball's speed of motion changes to a full stop when the ball collides with an outfielder's glove. In each of these collisions, energy is transferred and the motion of the ball is changed.

Wrap It Up!

1. **Name** What two features of motion can change when energy is transferred to an object?

2. **Apply** Describe what would happen if you transferred energy to a motionless soccer ball by kicking it.

3. **Infer** Does the ball's energy increase or decrease when its motion has stopped?

Investigate

Collisions

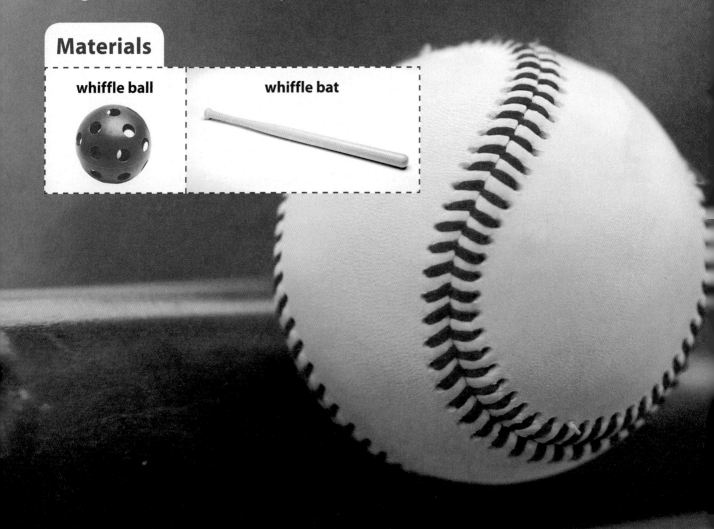

? **How does changing the speed of colliding objects change the amount of energy transferred?**

The pitcher, the teams, the fans... everyone breathlessly watches the motion of a baseball after it makes contact with the bat. They are all eager to know how the ball's energy and motion will change. Will it fly right past the outfielder or collide with his glove and then change direction? In this investigation, you can explore how the amount of energy transferred to a ball changes when it collides with objects in different ways.

Materials

- whiffle ball
- whiffle bat

PE 4-PS3-3. Ask questions and predict outcomes about the changes in energy that occur when objects collide.

1 Kneel about 3 meters away from a partner. Have your partner roll the ball to you. Observe and record what happens when you catch the ball. What objects are colliding? Identify and record any changes in the speed or direction of these objects. Roll the ball back to your partner.

2 Hold the bat flat on the floor. Have your partner roll the ball so that it collides with the bat. Observe and record what happens to the objects' speed and direction as they collide. Roll the ball back to your partner.

3 Again, hold the bat flat on the floor. Have your partner roll the ball to you. While keeping the bat on the floor, slowly swing the bat at the ball. Observe and record what happens to the objects' speed and direction as they collide. Roll the ball back to your partner.

4 Repeat step 3, but this time, swing the bat at a slightly faster speed. Observe and record what happens to the objects' speed and direction as they collide.

Wrap It Up!

1. **Describe** How does catching a ball change its energy?

2. **Ask** Write your own question and answer about how the energy of a ball changes when it collides with an object.

3. **Predict** What might happen to the energy of the ball if you hit it harder?

Sounds of the Game

Whoosh... smack! When you attend a baseball game, energy is moving all around you in the form of sound. A ball whooshes through the air and smacks into the catcher's mitt. A roar erupts from the crowd as the umpire yells, "You're out!" All these sounds occur as energy is transferred through the air in rapid back-and-forth movements called **vibrations.**

When the moving ball smacks into a catcher's mitt and comes to a stop, its energy isn't lost. Some of the energy is changed, or **transformed,** into heat. The surrounding air gets heated, vibrates, and sound is produced.

When the ball hits the catcher's mitt, the energy of its motion is transformed into heat and sound.

DCI PS3.A: Definitions of Energy. Energy can be moved from place to place by moving objects or through sound, light, or electric currents. (4-PS3-2), (4-PS3-3)

DCI PS3.B: Conservation of Energy and Energy Transfer. Energy is present whenever there are moving objects, sound, light, or heat. When objects collide, energy can be transferred from one object to another, thereby changing their motion. In such collisions, some energy is typically also transferred to the surrounding air; as a result, the air gets heated and sound is produced. (4-PS3-2), (4-PS3-3)

As the ball slams into the catcher's mitt, tiny air particles around the mitt are moved. These air particles push the air particles around them, and so on. As a result, waves of air vibrations quickly spread throughout the entire stadium. When the vibrations reach a person's eardrums, the person hears the sound.

When the player catches the ball, some of the ball's energy of motion will transform to sound energy. Some of the player's energy of motion will change to sound when he hits the ground, too.

Wrap It Up! My Science Notebook

1. **Recall** How does energy of motion transform into sound?

2. **Infer** Explain why covering your ears with your hands reduces the sound you hear.

3. **Infer** Why might a catcher's mitt feel warm after he catches a ball?

Investigate

Sound

? What evidence can you observe that sound transfers energy?

Your eardrums detect small vibrations, but some loud noises produce vibrations you can feel with the rest of your body. For example, when music at a baseball park plays over a loudspeaker, you might feel the vibrations. Sometimes you can see the vibrations produced by sound. In this investigation, you can observe the effect of sound on grains of salt.

Materials

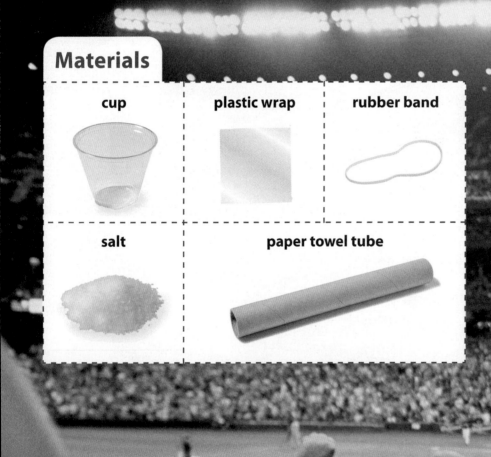

- cup
- plastic wrap
- rubber band
- salt
- paper towel tube

1 Use a rubber band to attach plastic wrap over the opening of the cup. Sprinkle salt in the center of the plastic wrap.

2 Hold the paper towel tube so that it is directed at the salt but not touching it. Predict what will happen when you speak into the tube. Record your prediction.

3 Speak softly into the paper towel tube. Observe what happens to the salt grains. Record your observations.

4 Repeat step 3, but this time speak loudly into the tube. Observe what happens to the salt grains. Record your observations.

Explore on Your Own

How can you produce another kind of sound that causes the salt to vibrate? Plan and carry out your own investigation. Record your observations. Compare the results of your investigations.

Wrap It Up!

1. **Observe** Did your observations match your prediction? Explain.

2. **Compare and Contrast** Describe what happened to the salt grains when you spoke into the tube loudly, then softly. Use the words *energy transfer* in your answer.

3. **Give Evidence** Using your observations as evidence, explain how energy can be transferred from place to place by sound.

33

The Sun's Light

What can transfer energy faster than the motion of a fastball or the vibrations of sound? Light. Light is energy that you can see. In about eight minutes, or the time it takes you to find your seat in a baseball stadium and sit down, light energy has already traveled all the way from the sun to your eyes. That's about 149,600,000 kilometers (92,960,000 miles)!

Light transfers energy from the sun through space. When the sun's light energy reaches Earth, that energy is transferred to countless objects and transformed in different ways. Objects sitting in sunlight become warm as the light is transformed into heat. Light also bounces off objects in our surroundings. When the light enters our eyes, it allows us to see.

The immense energy of the sun comes from chemical reactions that occur deep inside the sun's core.

DCI PS3.A: Definitions of Energy. Energy can be moved from place to place by moving objects or through sound, light, or electric currents. (4-PS3-2), (4-PS3-3)
DCI PS3.B: Conservation of Energy and Energy Transfer. Energy is present whenever there are moving objects, sound, light, or heat. When objects collide, energy can be transferred from one object to another, thereby changing their motion. In such collisions, some energy is typically also transferred to the surrounding air; as a result, the air gets heated and sound is produced. (4-PS3-2), (4-PS3-3) • Light also transfers energy from place to place. (4-PS3-2)

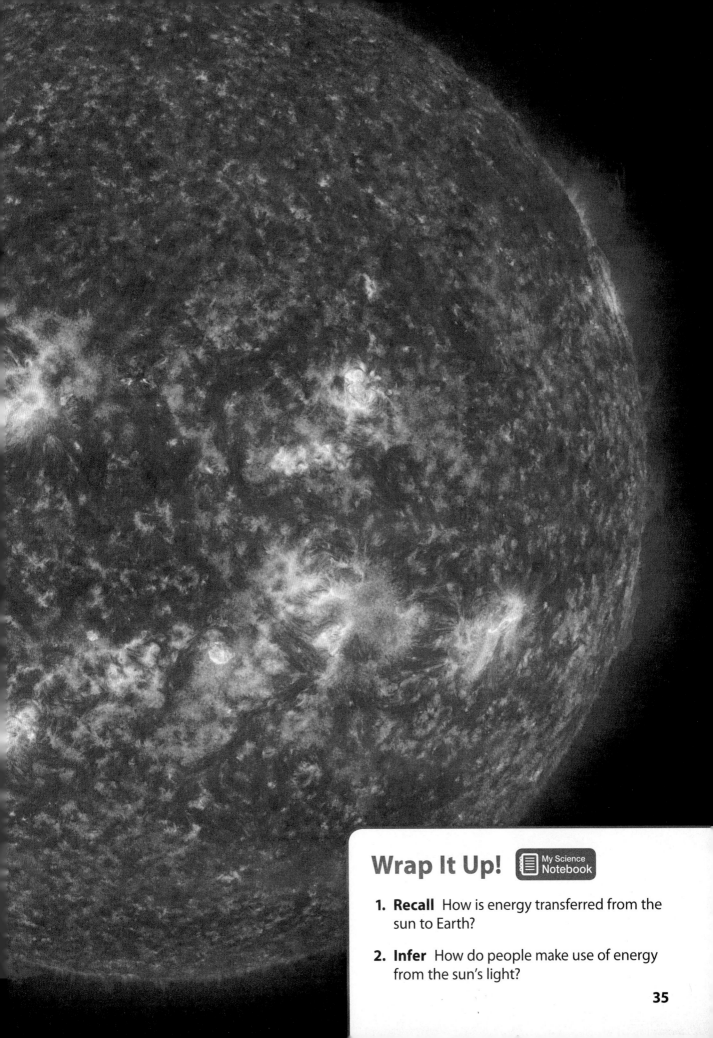

Wrap It Up!

1. **Recall** How is energy transferred from the sun to Earth?

2. **Infer** How do people make use of energy from the sun's light?

Investigate
Light

? What evidence can you observe that light transfers energy?

Just like sound energy can change the motion of salt grains, light energy can change objects. When direct sunlight strikes a piece of construction paper, the light causes a change in the paper. In this investigation, you can observe the effect of light energy on the color of construction paper.

Materials

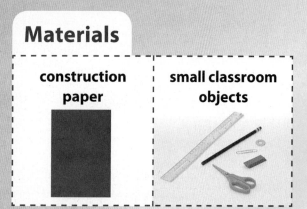

construction paper

small classroom objects

1 Place a piece of construction paper on a flat surface in a sunny spot.

2 Select a few objects to place on the construction paper.

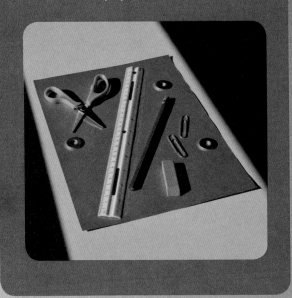

3 Predict what will happen to the color of the paper after it has been exposed to sunlight. Record your prediction.

4 Leave the paper alone until it has received a full hour or more of sunlight. Remove the objects and record your observations.

Explore on Your Own

What would happen if light-colored and dark-colored construction paper were exposed to sunlight? Plan and carry out your own investigation. Record your observations. Compare the results of your investigations.

Wrap It Up!

1. **Predict** Did your results support your prediction? Explain.

2. **Infer** What do your results show about how the transfer of light energy can be blocked?

3. **Give Evidence** Using your observations as evidence, explain how that energy can be transferred from place to place by light.

37

Stories in Science

Captured by Light

Pages from Anna Atkins's book reveal the fine details she was able to capture in her cyanotypes.

Anna made careful scientific drawings, like this one of a shell. Still, she felt a photograph could capture much more detail than a drawing.

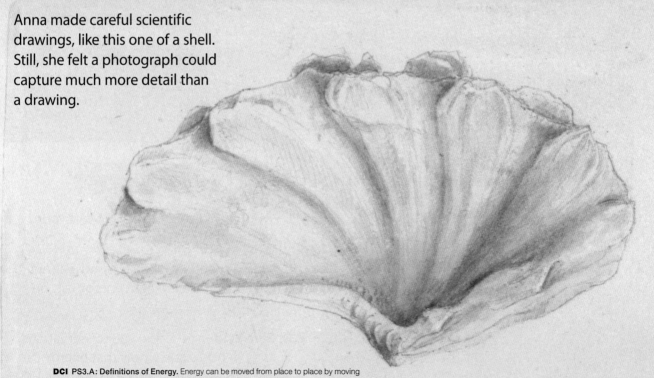

Anna Atkins combined her love of nature and art to study science in a new way.

In 1843, Anna Atkins became the first person to publish a book of photos. Many people name Anna as the first woman photographer. The details captured in Anna's photos gave scientists a new way to study nature.

Anna was born in England in 1799. Her mother died when she was a year old. Her father, a scientist, raised her. He taught Anna science at a time when most women did not get an education. Anna loved science and art. She did all the drawings in a book about shells that her father wrote.

Anna had a great interest in seaweeds and other algae. About 20 years after she had drawn shells for her father, Anna made her first photos of algae. She did not use a camera like we use today. Instead, she used a method invented by a family friend, John Herschel.

John had invented cyanotype as a way to print images on paper. He coated paper with a chemical and placed it in sunlight. The sun's energy was transferred to the paper. It caused a change in the chemical that turned the paper bright blue, or cyan. John was using his method to make copies of his notes. Anna had a different idea.

Working in the dark, Anna placed seaweed on coated paper. She moved it to a sunny spot. Sunlight turned most of the paper cyan. The light did not reach parts of the paper covered by seaweed. Those parts remained white. Anna had made a white image of seaweed surrounded by blue.

In time, Anna published about 200 of her photos of algae in a book. It was the first of its kind. The photos were useful as well as beautiful. They offered a whole new level of detail for the study of natural objects.

Wrap It Up!

1. **Explain** How is energy from sunlight important to making cyanotypes?

2. **Analyze** How did the process you used in the investigation "Light" compare to the method Anna used?

3. **Apply** What life experiences helped lead to Anna's success?

Heat It Up!

Heat is another way for energy to move from place to place. Energy is always present whenever heat is present. The particles that make up matter are always vibrating or moving. The energy of the moving particles is **thermal energy.** The faster the particles are moving, the more thermal energy the matter has. Heat is the transfer of thermal energy from a warmer object to a cooler object.

This camper can keep his drink warm using thermal energy from the campfire.

DCI PS3.A: Definitions of Energy. The faster a given object is moving, the more energy it possesses. (4-PS3-1)
DCI PS3.B: Conservation of Energy and Energy Transfer. Energy is present whenever there are moving objects, sound, light, or heat. When objects collide, energy can be transferred from one object to another, thereby changing their motion. In such collisions, some energy is typically also transferred to the surrounding air; as a result, the air gets heated and sound is produced. (4-PS3-2), (4-PS3-3)

A campfire has a lot of thermal energy. The camper kneels next to the campfire on a cool, dark night. Some of the energy of the campfire is transferred to the camper and helps him keep warm.

Heat can be used in many other ways. Thermal energy in buildings keeps us warm. You probably dry your clothes using thermal energy from a clothes dryer.

This popcorn maker heats popcorn seeds until they pop out of the hard outer shells.

Warm air from a hair dryer causes water from wet hair to evaporate.

Wrap It Up!

1. **Define** What is heat?
2. **Apply** In what ways do you use thermal energy every day?
3. **Relate** Besides heat, what other forms of energy are transferred by the campfire?

Investigate

Heat

? What evidence can you observe that heat transfers energy from place to place?

Warm air can transfer thermal energy to butter and melt it. That's why butter left out on a hot day turns into a liquid puddle! In this investigation, you can observe how a metal spoon can transfer thermal energy between water and butter.

Materials

| 3 cups with water of different temperatures | 3 metal spoons | 3 small dabs of butter |

PE 4-PS3-2. Make observations to provide evidence that energy can be transferred from place to place by sound, light, heat, and electric currents.

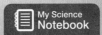 My Science Notebook

1 Place three cups of water on your desk. Record descriptions of the temperatures of the water in your science notebook.

2 Place a dab of butter on the stem of each spoon. Predict which dab of butter will melt first after the spoons are placed in the cups of water. Record your predictions.

3 Work with your group to place the spoons with butter in the cups of water. Be sure you put the spoons in the water at the same time.

4 Observe the butter every minute for 10 minutes. Record your observations.

Explore on Your Own

How would using a spoon made of a different material affect how thermal energy is transferred? Plan and carry out your own investigation. Record your observations. Compare the results of your investigations.

Wrap It Up!

1. **Observe** List the cups of water in order from least to most thermal energy.

2. **Predict** Did your results support your predictions? Explain.

3. **Give Evidence** How did the amount of thermal energy in the cups affect the melting of the butter? Use evidence from your observations in your explanation.

It's Electric

You probably know that to get the light bulb in a lamp to light, you need to plug the cord into an electrical outlet and turn on the switch. To produce light, the lamp uses electricity, or **electrical energy.** Electrical energy is the energy of moving charged particles. When electrical energy is transferred through wires, it is called **electric current.** Look at the photograph of the many carnival lights. What makes all these lights glow and the rides move? Electricity!

Electric current is a useful way to transfer energy from place to place. Energy from a faraway power plant can be moved through wires to all the buildings in a town. There it can be transformed into light, heat, sound, and motion.

At a carnival, electric current is transformed into colorful lights, festive sounds, and moving rides.

DCI PS3.A: Definitions of Energy. Energy can be moved from place to place by moving objects or through sound, light, or electric currents. (4-PS3-2), (4-PS3-3)
DCI PS3.B: Conservation of Energy and Energy Transfer. Energy can also be transferred from place to place by electric currents, which can then be used locally to produce motion, sound, heat, or light. The currents may have been produced to begin with by transforming the energy of motion into electrical energy. (4-PS3-2), (4-PS3-4)
CCC Energy and Matter. Energy can be transferred in various ways and between objects. (4-PS3-2)

Electric current transfers energy to a waffle iron, where it is transformed into heat.

The electric guitar is attached to an amplifier. The amplifier makes the sound from the guitar louder.

Wrap It Up!

1. **Define** What is electric current?

2. **Recall** Give examples of how electrical energy can be used to produce light, heat, sound, and motion.

3. **Apply** You have learned about some uses of electrical energy. What other ways do you use electrical energy every day?

Electric Circuits

The colorful lights in the photograph run on electricity. The wires and light bulbs make up an **electric circuit.** An electric circuit is a complete path through which electric current can pass. A battery and a power plant are sources of electricity for circuits.

A battery has stored energy. The battery has a positive end and a negative end. If you attach a wire from one end of the battery to the other, you produce a complete path through which an electric current can pass. For the current to flow, the circuit must be complete. If the wire is not connected to both ends of the battery, the energy cannot move through the circuit.

For these lights to shine, they must be connected to a source of electricity in a complete circuit.

DCI PS3.A: Definitions of Energy. Energy can be moved from place to place by moving objects or through sound, light, or electric currents. (4-PS3-2), (4-PS3-3)
DCI PS3.B: Conservation of Energy and Energy Transfer. Energy can also be transferred from place to place by electric currents, which can then be used locally to produce motion, sound, heat, or light. The currents may have been produced to begin with by transforming the energy of motion into electrical energy. (4-PS3-2), (4-PS3-4)

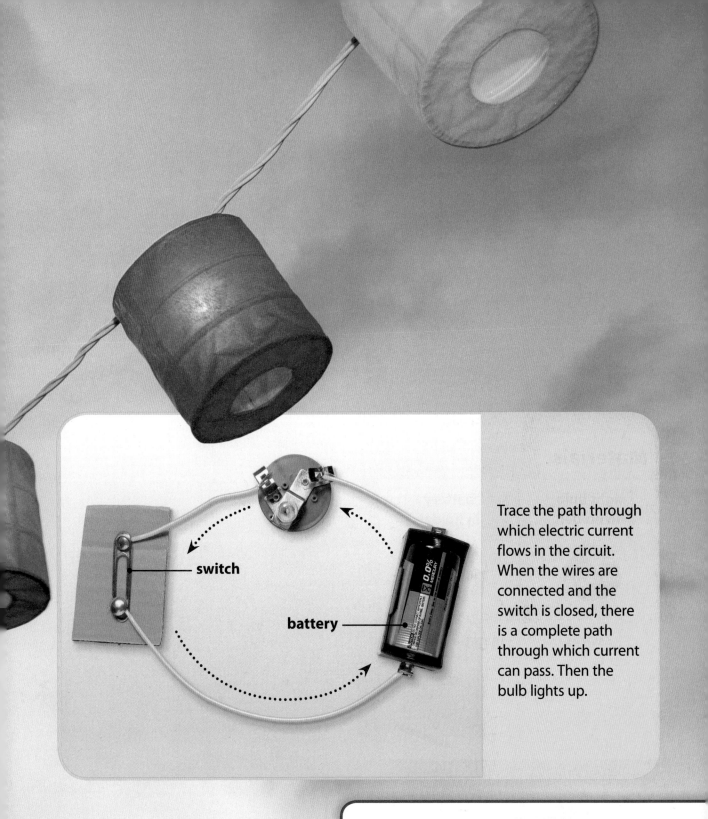

Trace the path through which electric current flows in the circuit. When the wires are connected and the switch is closed, there is a complete path through which current can pass. Then the bulb lights up.

Wrap It Up!

1. **Describe** How does an electric circuit transfer energy?

2. **Predict** What happens to the flow of electric current when you flip a light switch off? Explain.

47

Investigate
Electric Circuits

? **Which materials can complete an electric circuit?**

For electricity to flow, a circuit must be complete. Then the electric current can flow and light a bulb. In this investigation, you can observe which materials can complete an electric circuit.

Materials

light bulb in holder	battery in holder
3 wires	materials to test

1 Attach wires to a battery holder and to a bulb holder as shown.

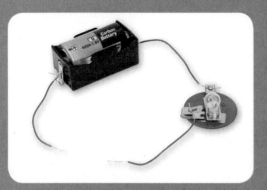

2 Notice that there are two wire ends that are not connected. Touch the ends of the wires together to make a complete circuit. Record your observations in your science notebook.

3 Predict whether a rubber band will complete the circuit if you touch the wire ends to it. Record your prediction. Touch the ends of the wires to the rubber band. Record your observations.

4 Predict whether other materials will complete the circuit. Record your predictions. Test the materials. Record your observations.

Explore on Your Own

What would happen if you added a second light to the electric circuit? Plan and carry out your own investigation. Record your observations. Compare the results of your investigations.

Wrap It Up!

1. **Predict** Did your results support your predictions? Explain.

2. **Describe** What materials are needed to build a complete circuit? How are the materials used to complete the circuit?

3. **Conclude** How can energy be transferred from place to place by electric currents? Use evidence from the activity in your answer.

Spin It!

Where does the electricity you use in your home come from? It must be generated using other forms of energy. Energy can't be created or destroyed. But energy can be changed, or transformed, from one form into another. One source of energy is wind. The moving air causes the wing-like blades of a wind turbine to turn around and around. The rotating blades have **energy of motion** that can be transformed into electricity.

DCI PS3.B: Conservation of Energy and Energy Transfer. Energy can also be transferred from place to place by electric currents, which can then be used locally to produce motion, sound, heat, or light. The currents may have been produced to begin with by transforming the energy of motion into electrical energy. (4-PS3-2), (4-PS3-4)

The energy from the moving blades of a wind turbine runs a generator. The generator transforms the energy of motion into electricity. People can use this electricity in homes, schools, and offices.

Wind turbines on a wind farm capture energy of motion of moving air.

Water turbines transform the energy from moving water in the ocean into electricity people can use.

Wrap It Up!

1. **Recall** Describe two sources of energy of motion.

2. **Explain** How is the energy from the moving blades of the turbines used to produce electricity?

51

STEM
SCIENCE
TECHNOLOGY
ENGINEERING
MATH

SPACE STATION PROJECT

Design a Collision Shield

As many as 100 million pieces of space junk are orbiting Earth. This junk has broken off for one reason or another from machines sent into space. Most space junk is very tiny—as tiny as flakes of paint. But space junk travels at high speeds. Even tiny objects can be dangerous to a spacecraft if they collide with it.

The International Space Station cannot steer around every tiny bit of space junk as it travels in space. Engineers have had to design shields to cover many of its surfaces. One type of shield is designed with layers that absorb, or spread out, energy from collisions. The layers may also break apart or even melt down an object after it hits.

Engineers also design equipment to reduce impacts of collisions on Earth. Bumpers protect cars. Helmets protect people's heads. You have even read about buildings designed to absorb the shock of earthquakes. Now it's your turn. You will work with your team to design a shield that protects an object during a collision.

PE 3-5-ETS1-3. Plan and carry out fair tests in which variables are controlled and failure points are considered to identify aspects of a model or prototype that can be improved.

PE 4-PS3-1. Use evidence to construct an explanation relating the speed of an object to the energy of that object.

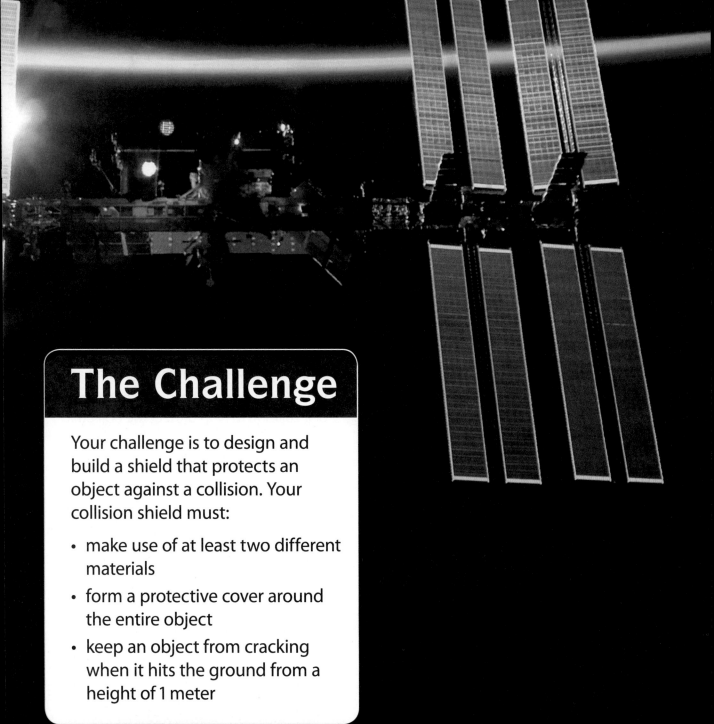

The Challenge

Your challenge is to design and build a shield that protects an object against a collision. Your collision shield must:

- make use of at least two different materials
- form a protective cover around the entire object
- keep an object from cracking when it hits the ground from a height of 1 meter

STEM
SPACE STATION PROJECT
(continued)

These bikers wear protective gear, including helmets and padded clothing. How can this gear shield their bodies if they fall?

1 Define the problem.

Imagine dropping a hardboiled egg from a height of 1 meter. Could you prevent its shell from cracking when it collides with the ground? This is the problem your team will work on. How can a shield help solve the problem? Write the problem in your science notebook.

Criteria tell what your solution needs to do to be successful. Look back at the challenge box on the previous page. These are your criteria. Write your criteria in your science notebook.

Your teacher will show you materials to select from when designing your shield. He or she will tell you how much time you have to work on a design. You cannot use any other materials, space, or time. These are the constraints of your solution. List the constraints in your science notebook.

2 Find a solution.

Look at the materials. How can you use them to make a shield? Think about these questions:

- Which materials best fit the object's shape?
- Which materials will best reduce the energy of a collision to help protect the object?
- Can any of the materials be changed or combined for greater protection?

Be creative! Sketch your design, and present it to your team. Discuss each possible solution. Choose the one that will best meet the criteria.

Draw your final design. Explain why your design will be successful and why you chose the materials you are using. Have your teacher approve your design.

3 Test your solution.

Build your shield. Record notes about your work in your science notebook. Make a plan to test your design. Your test will tell how well your collision shield meets the criteria. Decide which evidence will prove that your solution is successful.

Remember that your egg will be dropped from the height of 1 meter. To make your test fair, this variable should not change. You should also carry out several trials to collect data.

Have your teacher approve your plan. Then carry out your tests. Record information about all your tests and results in your science notebook.

Discuss the results with your team. Did your collision shield absorb enough energy? Why or why not? Talk about the other criteria. Did your design solve the problem?

4 Refine or change your solution.

Discuss ways you could improve your design. Do you need to change any of the materials you are using? Draw your ideas.

Test your collision shield the same way you did in Step 3. Hold your egg 1 meter above the ground for each trial. Did your changes improve your shield? How do you know?

Present your team's design to the class. Explain what worked well and what did not. Tell about the criteria and whether your collision shield met them. Explain how you tested your solution. Answer questions about your design. Ask questions about the other teams' designs.

How could your design be used in everyday life? How could you redesign your collision shield to be able to use it in other ways? Talk about your ideas with your team. Record your ideas in your science notebook.

NATIONAL GEOGRAPHIC

Think Like an Engineer
Case Study

Finding Solutions to Energy Problems

Problem

How can people meet their energy needs in places with limited resources?

T.H. Culhane helps people all over the world meet their energy needs. T.H. is an expert in finding solutions to energy problems. He got his start studying people in the rain forests of Borneo. The people had few resources available, yet they thrived in their environment. T.H. thought, "How can we use the idea of working with limited resources to do the same thing in the cities?" Let's talk with T.H. about some of his environmental and health solutions.

NGL Science Why did you go to rain forests?

T.H. Culhane I learned that people who live in the rain forest get all of their energy, all their food, and all their ecosystem services—recycling their wastes back into energy and food—from the forest. I went to live with them to learn how they did it.

NGL Science Your work takes you all over the world. You always try to find sustainable energy solutions. What does that mean?

DCI ETS1.A: Defining Engineering Problems. Possible solutions to a problem are limited by available materials and resources (constraints). The success of a designed solution is determined by considering the desired features of a solution (criteria). Different proposals for solutions can be compared on the basis of how well each one meets the specified criteria for success or how well each takes the constraints into account. (secondary to 4-PS3-4)
NS Science Is a Human Endeavor. Most scientists and engineers work in teams. (4-PS3-4)
NS Science Is a Human Endeavor. Science affects everyday life. (4-PS3-4)
CETS Influence of Engineering, Technology, and Science on Society and the Natural World. Engineers improve existing technologies or develop new ones. (4-PS3-4)

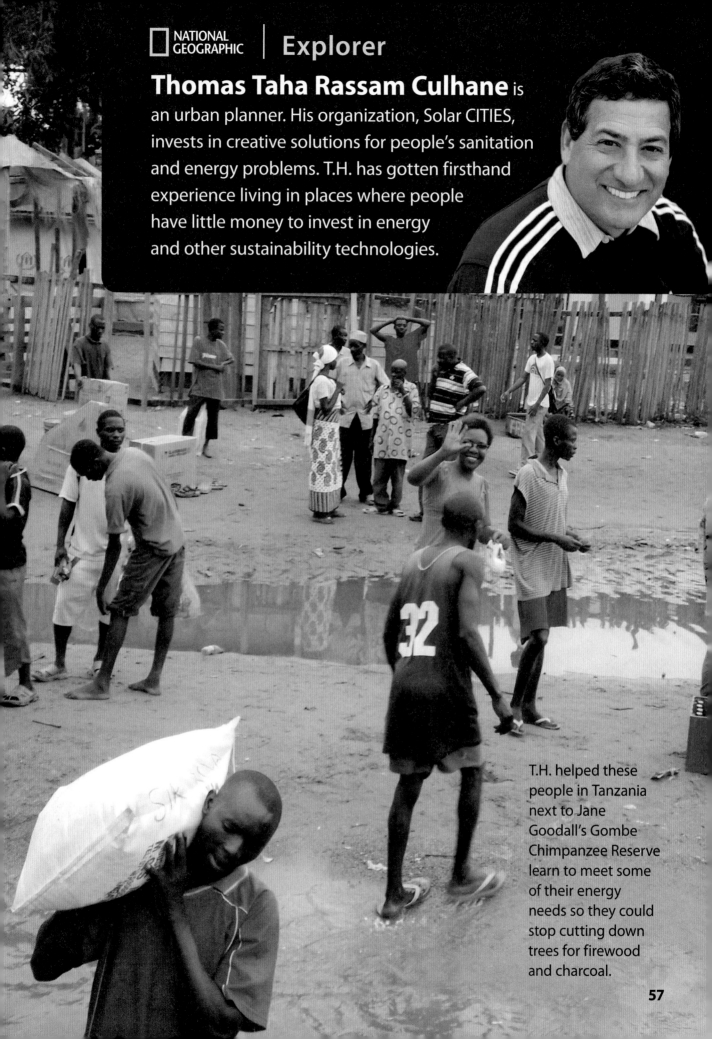

NATIONAL GEOGRAPHIC | Explorer

Thomas Taha Rassam Culhane is an urban planner. His organization, Solar CITIES, invests in creative solutions for people's sanitation and energy problems. T.H. has gotten firsthand experience living in places where people have little money to invest in energy and other sustainability technologies.

T.H. helped these people in Tanzania next to Jane Goodall's Gombe Chimpanzee Reserve learn to meet some of their energy needs so they could stop cutting down trees for firewood and charcoal.

Think Like an Engineer
Case Study (continued)

T.H. Culhane *Sustainable* means doing something in such a way that you can keep doing it. For example: You can save some of your garden seeds from this year. Then plant the seeds next year. You can do this year after year. You'll never run out of seeds. That's a sustainable way to grow a garden.

NGL Science What energy problems have you identified?

T.H. Culhane One: How can we get enough energy? Most energy people use originally comes from the sun. We have to transform energy from the sun into a form we can use. Two: How can we get rid of wastes without harming the environment?

Solution

NGL Science How have you helped people solve their energy problems in places with limited resources?

T.H. Culhane The idea is quite simple. Most energy people use originally comes from the sun. For example, the sun's energy makes plants grow. When animals eat the plants, they create waste from what they eat. We can use the waste for energy. We just need to transform the energy into a form we can use.

T.H. Culhane explains an energy solution to village leaders in Tanzania, Africa. **1)** A holding tank is installed. **2)** Animal wastes are added to the tank. **3)** The expandable tank that captures and holds the methane gas is installed.

T.H. Culhane and villagers add ground up food scraps, dead flowers, and other organic wastes to the biodigester tank. This dramatically increases gas production.

Think Like an Engineer
Case Study (continued)

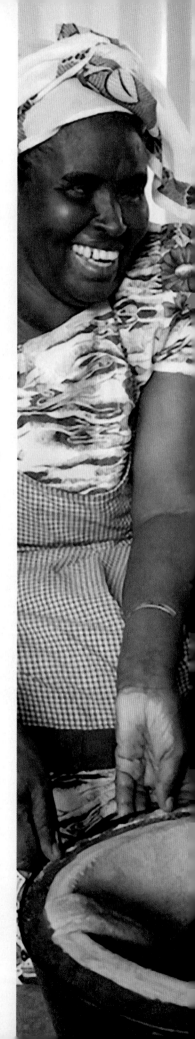

NGL Science How do we transform energy into forms we can use?

T.H. Culhane We have developed a simple device called a "home-scale" biodigester. The biodigester can transform energy in household wastes into energy and nutritious fertilizer we can use. To minimize energy use, we frequently use recycled materials to build this device.

NGL Science How did you get useful energy from wastes?

T.H. Culhane The biodigester transforms organic wastes. We can grind up our food scraps. Then the chopped up food gets added to the biodigester. So does human waste from the toilet. Bacteria feed on the wastes. As they feed, the bacteria produce methane gas. We burn the methane gas on the stove for cooking. You can even burn the gas in a generator to produce electricity!

NGL Science Does the biodigester produce anything besides gas?

T.H. Culhane Yes, the process also produces a rich liquid fertilizer. We use it in gardens.

NGL Science How did you get the idea of turning garbage into energy?

T.H. Culhane It came from years spent working on my Ph.D. with the Zabaleen trash pickers community of Cairo, Egypt, who make their living from recycling garbage. Getting rid of wastes without harming the environment is important. Some cultures understand that everything has a use. Food wastes and toilet wastes are useful. They're forms of energy. So a problem is also a solution!

T.H. Culhane and the cook in a school in one of Nairobi's most crowded urban slums prepare to use a new stove fueled by biogas that eliminates the smoke, health, and environmental problems associated with charcoal.

Wrap It Up!

1. **Explain** What constraints, or limits, did T.H. Culhane have to keep in mind when designing his energy solutions?

2. **Explain** How does T.H. Culhane's energy solution make use of available materials?

3. **Analyze** Why do you think T.H. Culhane named his energy device a biodigester?

61

NATIONAL GEOGRAPHIC | Think Like an Engineer

Design, Test, and Refine a Device

You've read how T.H. Culhane applied scientific ideas to design his biodigester. As he made and refined his design, he accounted for the lack of resources in the areas his device would be used. Now it's your turn. Imagine that you are working to find an energy solution for people living in remote mountain communities without access to electricity. The current solution of using wood fires to cook food is not sustainable. Wood fires use up forest resources and produce harmful smoke in people's homes.

1. **Define the problem.**

    Solar ovens convert sunlight energy into heat. What problem would solar ovens solve for the mountain communities? To be successful, what must your oven be able to do? How will you know it works? These are the criteria your oven must meet. Next, think about the constraints for your solution. Remember, the people using this oven will need to build it far away from a hardware store. You must use simple, inexpensive materials. The oven must be easy to construct. It must also be safe to use.

2. **Find a solution.**

    Study several solar oven designs, and then draw and label your own design. List the beginning and final forms of energy. Explain how your oven will transfer energy from one form to another. Describe how your design will meet the criteria for a successful solution. Create a prototype of your solar oven using available materials.

PE 4-PS3-4. Apply scientific ideas to design, test, and refine a device that converts energy from one form to another.
PE 3-5-ETS1-1. Define a simple design problem reflecting a need or a want that includes specified criteria for success and constraints on materials, time, or cost.
PE 3-5-ETS1-3. Plan and carry out fair tests in which variables are controlled and failure points are considered to identify aspects of a model or prototype that can be improved.

This woman in the African country of Mali prepares food using a solar cooker.

Think Like an Engineer
(continued)

3. **Test your solution.**
   Decide how you will test your prototype. What measurements will you take to provide evidence that your solar oven works? What tools will you need to take those measurements? When your teacher has approved your plan and you have what you need, carry out your test. Record your observations.

4. **Refine or change your solution.**
   Analyze your findings. How could your prototype be improved? Refine your design and test it again. Record your changes.

5. **Compare your data.**
   Did your change improve the performance of your solar oven design? Do you need to make more refinements? Continue to refine your prototype until you are happy with its performance.

6. **Analyze and explain your results.**
   Explain the scientific basis of your design solution.

   - What forms of energy are involved in your design?
   - How does your oven transform one type of energy into another?
   - Which criteria does your design meet?
   - What evidence do you have to show that your prototype is successful?
   - How could your design be further improved?

7. **Present your results.**
   Present your arguments and your prototype to the class. Evaluate your classmates' prototypes using the same criteria you used to evaluate your own.

This woman in Chad, a country in northern central Africa, makes tea with water boiled by a solar oven. She did not have to gather firewood.

NATIONAL GEOGRAPHIC | **Think Like an Engineer**

Design, Test, and Refine a Device

A new board game is coming out soon. The game developers have asked you, an engineer, to design a buzzer for the game. The buzzer will need to be battery-powered and controlled by a switch. Your buzzer design must use materials from among those provided by your teacher, and be quick and low-cost to build.

1. **Define the problem.**

   You know you need to build a buzzer that can be turned on and off. You can use a circuit similar to the one you built in *Investigate Electric Circuits*, except a buzzer can be used in place of the light bulb. In order to turn the buzzer on and off, you will need to include a switch in your circuit. Look at the switch in the circuit on page 165 for hints on how to do that.

2. **Find a solution.**

   Think about the criteria your buzzer must meet. It must be battery-powered and controlled by a switch. Also think about the constraints of the design. Your buzzer must be able to be built quickly and at low cost. It must be safe to use. Study the materials available to you. Which ones will you use to build your prototype? Draw and label a diagram of your prototype. Explain how you think your prototype will work.

PE 4-PS3-4. Apply scientific ideas to design, test, and refine a device that converts energy from one form to another.
PE 3-5-ETS1-1. Define a simple design problem reflecting a need or a want that includes specified criteria for success and constraints on materials, time, or cost.
PE 3-5-ETS1-3. Plan and carry out fair tests in which variables are controlled and failure points are considered to identify aspects of a model or prototype that can be improved.

3. **Test your solution.**
   Plan how you will test your circuit design. What evidence will you provide to show that your circuit design meets the criteria and constraints? After your teacher has approved your plan and you have what you need, carry out your test. Record your observations.

4. **Refine or change your solution.**
   Analyze your results. Did your prototype work? Are there changes you could make that would improve your prototype? Make the changes and retest. Keep refining and testing your prototype until you are sure it is the best it can be.

5. **Analyze and explain your results.**
   Explain the scientific basis of your design solution.

   - What forms of energy are involved in your design?
   - How does it transform one type of energy into another?
   - Refer to the criteria you identified earlier for a successful device. Which criteria does your design meet?
   - What evidence do you have to show that your prototype is successful?

6. **Present your results.**
   Present your arguments and your design to the class. Evaluate your classmates' designs using the same criteria you used to evaluate your own.

This game buzzer converts electrical energy to light and sound energy.

Nonrenewable Energy Resources

Inside power plants, fuel such as coal, oil, or natural gas is burned to heat water and create steam. The steam turns the blades of turbines, providing energy of motion. Generators transform the energy of motion into electricity. Power plants do not "produce energy," though that is a common expression. They convert the stored energy from fuel into a form of energy for practical use.

Coal, oil, and natural gas are fossil fuels. A **fossil fuel** is a source of energy that formed from the remains of plants and animals that lived millions of years ago. Pressure and heat over time caused the remains to change into fossil fuels. Fossil fuels are considered **nonrenewable energy resources** because they will eventually run out.

Natural gas is a gas form of fossil fuel. Natural gas is drilled from the ground.

Coal is a fossil fuel that is mined from the ground using large machines.

Nuclear power plants use energy from materials such as uranium to produce electricity. Uranium is a nonrenewable energy resource mined from the ground.

Oil rigs remove oil and natural gas from under the ocean floor. These fossil fuels are being removed faster than nature can replace them.

Wrap It Up!

1. **List** Identify four nonrenewable energy resources.

2. **Identify** What type of energy is transformed into electricity in nuclear power plants?

3. **Interpret** What do people really mean when they say a power plant "produces" energy?

4. **Analyze** How might engineers work with scientists to design machines that harvest fossil fuels from Earth?

Renewable Energy Resources

Unlike fossil fuels and nuclear power, some sources of energy are **renewable energy resources.** Renewable energy resources will never run out. For example, **solar energy** and **wind energy** are renewable. No matter how much sunlight and wind we use to produce energy, there will always be more.

Hydroelectric dams capture the energy of motion from water in rivers. When water flows through the dam, it spins turbines inside the dam and generates electricity. Hydroelectric dams produce renewable energy.

Solar panels transform the sun's light and heat energy into electricity.

DCI ESS3.A: Natural Resources. Energy and fuels that humans use are derived from natural sources, and their use affects the environment in multiple ways. Some resources are renewable over time, and others are not. (4-ESS3-1)
CETS Influence of Engineering, Technology, and Science on Society and the Natural World. Over time, people's needs and wants change, as do their demands for new and improved technologies. (4-ESS3-1)

In windy locations, wind turbines transform energy from moving air into electricity. Most wind turbines are tall to catch the strong, steady winds high above the ground.

Wrap It Up!

1. **List** List three sources of renewable energy.

2. **Relate** How is electricity generation from wind and water similar to electricity generation from fossil fuels?

3. **Infer** Why might people be interested in switching to renewable energy resources?

Energy Resources and the Environment

People cannot produce energy to meet their needs without converting it from other sources. Scientists and engineers work on developing energy resources that can support society and the environment over a long time. Solar, wind, and water resources are renewable, and they are cleaner than fossil fuels. Still, no option is perfect. Even renewable resources affect the environment. Scientists and engineers are always looking for new sources of energy that have few disadvantages.

The chart on the next page gives some advantages and disadvantages of different energy resources.

A worker climbs a wind turbine at an offshore wind farm.

DCI ESS3.A: Natural Resources. Energy and fuels that humans use are derived from natural sources, and their use affects the environment in multiple ways. Some resources are renewable over time, and others are not. (4-ESS3-1)
DCI PS3.D: Energy in Chemical Processes and Everyday Life. The expression "produce energy" typically refers to the conversion of stored energy into a desired form for practical use. (4-PS3-4)
NS Science Is a Human Endeavor. Science affects everyday life. (4-PS3-4)

ADVANTAGES		DISADVANTAGES
Oil is relatively inexpensive because we already have the machines to obtain and use it.		Oil is nonrenewable, and burning oil produces air pollution. Oil spills harm ecosystems.
Coal is relatively inexpensive because we already have the machines to obtain and use it.		Coal is nonrenewable, and burning coal produces air pollution. Coal mining damages the land and pollutes water in the ground.
Natural gas burns more cleanly than the other fossil fuels.		Natural gas is nonrenewable, and drilling for natural gas can pollute air and water.
Nuclear energy generates electricity from nuclear reactions without polluting the air.		Uranium is nonrenewable. Wastes from nuclear energy plants can cause harm to living things, including people.
Solar energy is renewable and does not cause pollution.		Solar fields require large amounts of land, and their use depends on the availability of sunlight.
Wind energy is renewable and does not cause pollution.		Wind farms require large amounts of land and may harm bird populations. Wind does not blow steadily in all places.
Hydroelectric energy uses moving water. It is renewable and does not cause pollution.		Hydroelectric dams destroy river habitats and may disrupt fish populations.

Wrap It Up!

1. **Evaluate** Which resource do you think has the greatest disadvantage? Explain.

2. **Make Judgments** Is there such a thing as a "clean" fossil fuel? Explain.

3. **Analyze** What is one disadvantage of energy use that might affect you?

NATIONAL GEOGRAPHIC | Think Like a Scientist

Obtain and Combine Information

People's needs and wants change over time, as do their demands for new and improved technologies. More people live in the United States now than ever before, and our demand for electricity is greater than it ever has been. As we move toward the future, we need to look carefully at the technologies we use to produce energy. We must think about how well they meet our changing needs and how their use impacts our environment.

What are the environmental impacts of the different energy resources that we use? Conduct research to make a claim based on evidence.

1. **Carry out your research.**
   Look back to the chart in the previous lesson, "Energy Resources and the Environment." Work with a partner or group to research each energy resource in the chart. Collect more information from books, newspaper articles, and websites. Record the sources of all the information you obtain. Information without a source cannot be used.

PE 4-ESS3-1 Obtain and combine information to describe that energy and fuels are derived from natural resources and their uses affect the environment.
CETS Influence of Engineering, Technology, and Science on Society and the Natural World. Over time, people's needs and wants change, as do their demands for new and improved technologies. (4-ESS3-1)

The United States uses more than 19.5 million barrels of oil per day.

Use these objectives to guide your research.

- Describe how the energy resource is taken from a natural resource such as wind, sunlight, or coal.
- Identify the ways in which the energy resource is important to people.
- Identify the positive and negative environmental effects of using the energy resource.
- Find out whether or not technology can make the energy resource more helpful or less harmful.

2. **Organize and evaluate your research.** Organize the information you have collected about each energy resource. Categorize each one as renewable or nonrenewable.

- What evidence have you collected about the environmental effects of using each energy resource?
- What evidence is there about the role of technology in making it more helpful or less harmful to use? You may need to do more research to find the evidence you need to make a claim.

3. **Communicate information.** Use the information you have found to make an evidence-based claim about how using energy resources affects the environment. When you are ready, present your research to the class. Point out possible ways that technology may help protect Earth's resources.

Waves

Up and down, up and down, the surfer bobs. Energy is transferring through water beneath her. Energy travels through water in regular patterns of motion called **waves.** When water is disturbed, it moves up and down. This motion disturbs the water next to it, making the water there move up and down. This motion disturbs the water next to that, making the water in that place move up and down. Pretty soon the up-and-down motion has traveled far through the water. As each wave passes through the water beneath the surfer, the board she sits on rises and falls.

Water waves are a kind of transverse wave. Energy travels from side to side in a **transverse wave** and causes up-and-down movement of particles as it passes through them.

While the surfer waits for a wave that is tall enough to ride, smaller waves move her up and down without moving her forward. To "catch" a wave, the surfer paddles her board and begins moving. She matches the speed of a passing wave so that she is carried along.

DCI PS4.A: Wave Properties. Waves, which are regular patterns of motion, can be made in water by disturbing the surface. When waves move across the surface of deep water, the water goes up and down in place; there is no net motion in the direction of the wave except when the water meets the beach. (4-PS4-1)

As a water wave gets close to the beach, it gets tall. Some of the water "breaks" or curls over and falls toward the front of the wave. The surfer slides down the steep front of the wave.

Wrap It Up! My Science Notebook

1. **Describe** Describe how a wave travels across the ocean.

2. **Infer** As a wave moves into shore the water becomes more shallow and the wave slows down. What happens to its energy as a result?

Properties of Water Waves

Waves spread out in all directions from their source. If you have ever dropped a pebble into a pond and watched the ripples, you have seen a wave pattern. When you drop a pebble into a pond, it makes small up-and-down, or transverse, waves. A larger rock makes larger waves. The sizes of waves differ in height and in the distance between their highest points.

The distance between the highest point, or crest, of a wave and the middle point of the wave is called **amplitude.** Amplitude can also be measured from the lowest point, or trough, of the wave to the middle point of the wave.

Amplitude is one property of waves. **Wavelength** is another. Wavelength is the distance from one crest to the next crest, or one trough to the next trough.

DCI PS4.A: Wave Properties. Waves, which are regular patterns of motion, can be made in water by disturbing the surface. When waves move across the surface of deep water, the water goes up and down in place; there is no net motion in the direction of the wave except when the water meets the beach. (4-PS4-1) • Waves of the same type can differ in amplitude (height of the wave) and wavelength (spacing between wave peaks). (4-PS4-1)
CCC Patterns. Similarities and differences in patterns can be used to sort, classify and analyze simple rates of change for natural phenomena. (4-PS4-1)

Amplitude measures the height of crests and depth of troughs. Wavelength measures how far apart they are.

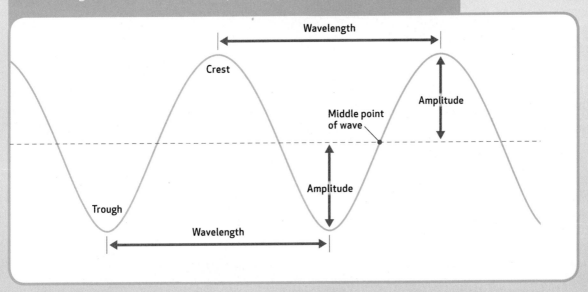

These water waves are not very tall or far apart. In other words, they have a low amplitude and a short wavelength.

Wrap It Up!

1. **Recall** What are amplitude and wavelength?

2. **Infer** Waves transfer energy of motion. Which transfers more energy: a wave with a high amplitude or a wave with a low amplitude? Explain.

Properties of Sound Waves

You have already learned that sound is a form of energy. Sound energy travels through air, water, soil, and other materials in the form of vibrations. Vibrations help carry the sound of this bird's song through the air in all directions.

Sound vibrations travel in a wave pattern. Sound waves are a type of longitudinal wave. In a **longitudinal wave,** the particles of a material move back and forth as they vibrate. The waves create areas in which particles press close together followed by areas in which particles spread farther apart.

Sound waves have the properties of wavelength and amplitude. Wavelength is related to a sound's **pitch,** or how high or low it sounds. Sounds with a higher pitch have shorter wavelengths. The waves vibrate more quickly.

The amplitude of a sound wave determines its **volume,** or loudness. The greater the amplitude, the greater the volume.

A yellow warbler makes high-pitched sounds.

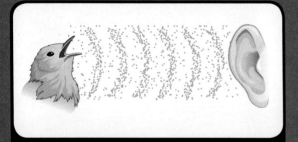

The energy of traveling sound waves causes air particles to press together and spread apart.

DCI PS4.A: Wave Properties. Waves of the same type can differ in amplitude (height of the wave) and wavelength (spacing between wave peaks). (4-PS4-1)
CCC Patterns. Similarities and differences in patterns can be used to sort, classify, and analyze simple rates of change for natural phenomena. (4-PS4-1)

SCIENCE in a SNAP

Make Waves

1 Place a coil on a flat surface. Stretch it just enough to find the middle. Tie a string around the coil's middle point.

2 With a partner, stretch and hold each end of the coil. Hold one end steady. Bounce the other end up and down. Identify the wave pattern you see.

3 Lay the coil on a flat surface and stretch it out. Hold one end steady. Push the other end in and out. Identify the wave pattern you see.

? How did the string and the coil move each time?

Wrap It Up! My Science Notebook

1. **Compare** How does the movement of particles in a sound wave compare to the movement of particles in a water wave?

2. **Apply** A bird makes a loud, high-pitched sound. What can you say about the amplitude and wavelength of the sound waves?

Investigate

How Waves Move Objects

? How can you model how waves cause objects to move?

For some people, the rise and fall of a boat's floor beneath their feet is enough to lull them to sleep. For others, it's enough to make them sick! In this investigation, you'll explore how waves transfer energy to objects, causing them to move.

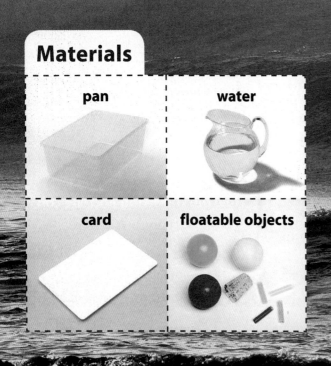

Materials

- pan
- water
- card
- floatable objects

 **My Science Notebook**

1 Fill the pan half full with water. Dip the short edge of the card down into the water at an angle. Keeping the card in the water, gently move the card up and down to produce a wave.

2 Select an object and place it in the water. Use the card to make a wave. Observe the wave and the motion of the object. Record your observations. Remove the object.

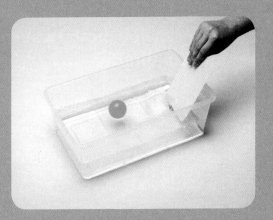

3 Now use the card to make a wave with a lower amplitude. Record your observations and what you did to make the wave.

4 Now use the card to make a wave with a shorter wavelength. Record your observations and what you did to make the wave.

5 Draw a diagram to model each wave you made in steps 3 and 4. Identify the amplitude and wavelength in each wave diagram. Write a sentence that compares the movement of an object caused by the waves in your diagram.

Wrap It Up! My Science Notebook

1. **Relate** How did your motions relate to the characteristics of each wave you made?

2. **Conclude** How did your model show how waves can cause objects to move?

Investigate

Wavelength and Amplitude

? **How can you model wavelength and amplitude?**

You have learned that wavelength and amplitude are two properties of waves. The water waves in the photograph are small and are close together. What would a diagram of these waves look like? On a very windy or stormy day, the waves might be large and far apart. What would a diagram of these waves look like?

In this investigation, you will use wire to model transverse waves with different wavelengths and amplitudes.

Materials

chenille stem   marker

1 Use the chenille stem to make a model of a wave. Describe its wavelength and amplitude. For example, it may have a long wavelength and low amplitude.

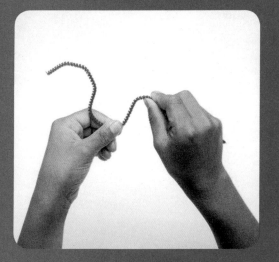

2 Draw and label a picture of your wave.

3 Keeping the amplitude the same, change the wavelength of your wave. Draw and label a picture of your wave.

4 Make a new wave with different properties. Draw and label your wave.

Explore on Your Own

How could you use your model to show and explain how a boat might ride the waves as the wavelength and amplitude change during a storm? Share your model and explanation with a partner.

Wrap It Up!

1. **Describe** Describe the properties of each wave you modeled.

2. **Analyze** How can two waves with the same wavelength differ?

85

Information Technology—GPS

When you look at a map on a smartphone, you can zoom in and out and even see views from the street. All of this is possible because the information has been put into digital code form, or **digitized.**

Digital maps usually have a little dot showing exactly where you are. How does the phone know where you are? The phone uses **GPS.** GPS stands for **global positioning system.** This system can locate anything with a GPS receiver.

Satellite

Receiver

Satellites in space **transmit,** or send, radio signals to a receiver on Earth, such as a smartphone. The phone receives the signals and uses them to pinpoint a location on a map.

DCI PS4.C: Information Technologies and Instrumentation. Digitized information can be transmitted over long distances without significant degradation. High-tech devices, such as computers or cell phones, can receive and decode information—convert it from digitized form to voice—and vice versa. (4-PS4-3)

GPS can be very useful in navigation, or finding your way.

Wrap It Up!

1. **Restate** What is GPS?

2. **Apply** List as many ways as you can think of to transmit a message. Which of those do you think use digital code?

Information Technology—Cell Phones

Radio signals, such as the ones used in GPS, can transmit over long distances. But radio signals aren't always stable. If you have ever tried talking to a friend with a radio walkie-talkie, you have probably noticed your friend's voice wasn't very clear.

A better way to communicate over long distances is with digital code. When you talk into a cell phone, the phone converts the sound of your voice into digitized form before transmitting it as a radio signal.

The radio signal is picked up by the nearest cell phone tower in your network. The tower redirects the signal through the network to your friend's phone. Your friend's phone converts the digitized information back into the sound of your voice.

Cell phones use digital code to maintain the sound quality of conversations over long distances.

DCI PS4.C: Information Technologies and Instrumentation. Digitized information can be transmitted over long distances without significant degradation. High-tech devices, such as computers or cell phones, can receive and decode information—convert it from digitized form to voice—and vice versa. (4-PS4-3)

Digital information remains stable over long distances. Digital information is coded using a pattern of 0s and 1s.

Wrap It Up!

1. **Sequence** Put the following sentences in the correct sequence: The digital information is transmitted through a radio network to the other phone. The phone converts the digital signal into sound. The phone converts sound into a digital signal.

2. **Compare** How do radio signals compare with digital signals for transmitting information over long distances?

Investigate
Use a Code

? How can you use patterns to transmit information?

People have been using codes to transmit information over long distances for years. In the 1800s, Samuel Morse invented a code that uses long and short signals. Originally transmitted with a telegraph, the dots and dashes represented faster or slower clicking signals. Morse code can also be transmitted with light using faster or slower flashes.

In this investigation, you can transmit your own message to a partner using Morse code.

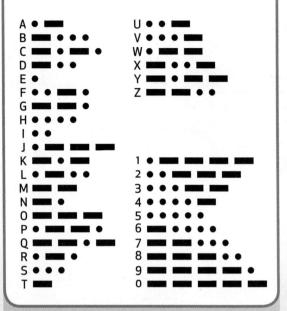

MORSE CODE

A ·—		U ··—	
B —···		V ···—	
C —·—·		W ·——	
D —··		X —··—	
E ·		Y —·——	
F ··—·		Z ——··	
G ——·			
H ····			
I ··			
J ·———			
K —·—		1 ·————	
L ·—··		2 ··———	
M ——		3 ···——	
N —·		4 ····—	
O ———		5 ·····	
P ·——·		6 —····	
Q ——·—		7 ——···	
R ·—·		8 ———··	
S ···		9 ————·	
T —		0 —————	

Each character in a word is represented by a series of dots and dashes. This is Morse code for SOS:

··· ——— ···

It stands for "save our ship" and was used by boats to signal the need for help. Now its use is not limited to boats.

Materials

flashlight

1 Study the Morse code key. Think of a message that is two to five words long. Record your message. Do not show your partner.

2 Use the flashlight to transmit your coded message to a partner. Remember to leave "spaces" between letters, and longer spaces between words.

3 Have your partner record your code and use the key to decode the message.

4 Compare the message you sent to the message your partner received. Record your observations.

To use a telegraph to communicate, both the sender and receiver must understand the code.

Wrap It Up!

1. **Explain** How well did your message transmit? Explain your results.

2. **Analyze** Which is easier to decode, a message sent in writing or a message in flashes of light? When might one work better than the other?

3. **Compare** How does Morse code compare with digital code as a way of transmitting information?

NATIONAL GEOGRAPHIC | **Think Like an Engineer**

Compare Multiple Solutions

You can find a bar code on the package of nearly every grocery item in your kitchen. Many smartphones have apps that translate bar codes into price information. Codes are all around you! But all codes don't require electronic devices to decode them.

You can invent your own code. Imagine that you are at summer camp and want to communicate with your friend in the next cabin, only a few feet away. Cell phones are not allowed. You both know how to use Morse code, but flashlights won't work because you can't see into each other's cabin. It's up to you to devise a way to communicate with each other.

Bar codes contain information that help stores keep track of the products they sell.

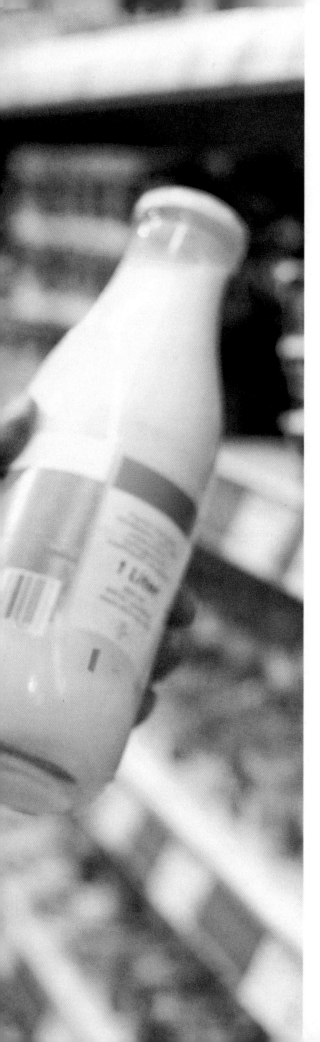

1. **Define the problem.**
   What problem needs to be solved? List the criteria for a successful solution. What constraints limit your choices? Constraints might include materials, distance the message must travel, or safety issues.

2. **Find a solution.**
   Study the materials available to you. Think about the challenge of transmitting Morse code. Which materials will allow you to turn a written message into a pattern of coded signals?

   Plan how you will test your idea. How will you simulate being in separate cabins? What message will you use to test your solution? What observations will you make? Record all your ideas in your science notebook.

3. **Test your solution.**
   What observations will you make? Once your teacher has approved your plan and you have what you need, carry out your test. Record your observations.

4. **Refine or change your solution.**
   Analyze your results. Did your solution work? Does it allow you to communicate clearly? How could your solution be further improved? Revise your solution and test it again.

5. **Share your solution.**
   Present your information technology solution to the class. Evaluate your classmates' solutions using the same criteria you used to evaluate your own. How many ways did your class come up with to communicate? How were the solutions similar and different? Which do you think is most effective?

93

Stories in Science

Code Talkers

Navajo Code Talkers used secret codes to send messages by radio during World War II.

This Congressional Gold Medal was awarded to the Navajo Code Talkers.

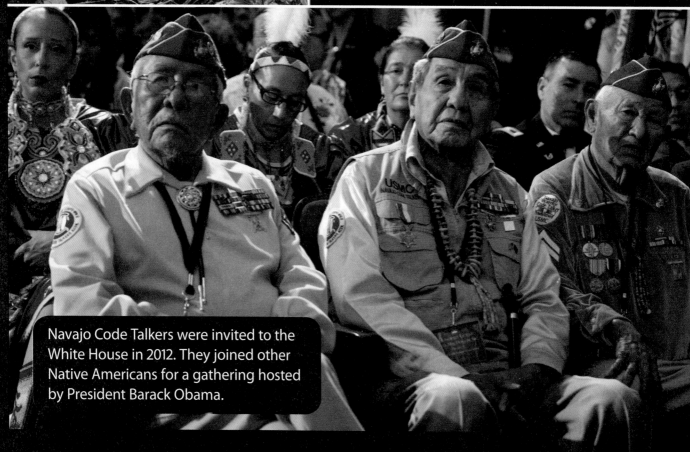

Navajo Code Talkers were invited to the White House in 2012. They joined other Native Americans for a gathering hosted by President Barack Obama.

"Diverse cultures can make a country richer and stronger." —Chester Nez, Navajo Code Talker

Codes help people talk to one another. Some codes, such as Morse code, are meant to be used by anyone. At one time, Morse code helped people communicate over long distances. Other codes are meant to be secret. All the countries that fought in World War II used secret codes. They coded messages sent by telephone and radio, so enemies would not know their plans. Each country also tried to decipher, or break, the others' codes.

The United States was one of the countries that fought in World War II. The Marines needed a code that other nations couldn't break. The Marines turned to 29 Navajo men to develop this code. Basing a code on the Navajo language was smart. The Navajo did not write their language down. Very few people who were not Navajo could speak it.

Making the code wasn't easy. The Code Talkers, as they were called, had no words in their own language for some of the weapons they had to talk about. For example, they had never seen most of the planes the Marines used. But the planes reminded them of birds. So the Code Talkers used Navajo words for different birds to stand for different kinds of planes.

Several Navajo words were chosen to stand for each letter. Using just one word for each letter, or 26 words, would have been too simple. The Navajo words for ant, apple, and axe stood for the letter *A*. In Navajo, none of these words begins with an *A* sound.

About 400 Navajo Code Talkers protected secrets during the war. Their codes were never broken. Their role remained top secret until 1968, more than 20 years after the war ended. In 2000, President Bill Clinton signed a law that awarded the original 29 Navajo Code Talkers the Congressional Gold Medal.

Wrap It Up!

1. **Explain** Why was the Navajo language a good choice for developing a code?

2. **Describe** How did the Navajo choose words to talk about ideas or objects that were new to them?

STEM
ENGINEERING PROJECT

SCIENCE
TECHNOLOGY
ENGINEERING
MATH

Design a Wind Instrument

WHEEEE ooooo EEEEE ooooo EEEEE. The pulsing wail of a siren fills the air. The loud sound warns everyone that a fire truck is coming. The high amplitude and changing wavelengths of the sound send a signal: "Watch out!" Drivers know to pull their cars to the side of the road. They let the fire truck safely pass by.

In this engineering project, you and your team will use what you have learned about sound to design a wind instrument. When you blow on a wind instrument, the air inside it vibrates, producing a sound. Then you will design a way to use your instrument to communicate. You will send a coded pattern of sounds that tell a teammate how to respond.

PE 4-PS4-3. Generate and compare multiple solutions that use patterns to transfer information.
PE 3-5-ETS1-2. Generate and compare multiple possible solutions to a problem based on how well each is likely to meet the criteria and constraints of the problem.

The Challenge

Your challenge is to design and build a wind instrument and use it to send a message. Your wind instrument must:

- produce sound when you blow into or across it
- make sounds of at least three different pitches
- send a coded pattern of sounds to communicate

Fire trucks and other emergency vehicles must travel quickly. They use sirens to clear traffic.

STEM
ENGINEERING PROJECT
(continued)

A band plays various brass instruments while on parade in Brazil. Their tube-shaped instruments produce sound as the musicians blow into them.

1 Define the problem.

Think about the problem you are solving. What does your wind instrument need to do? These things are the criteria of the problem. Criteria tell you if your design is successful.

There are also constraints, or limits to your design. Your teacher will give you materials to build your wind instrument. You cannot use any other materials. Your instrument must be a wind instrument. You cannot make an instrument that makes sounds when you tap on it or pluck a string. The message you send must be one your teacher gives you.

Write the problem in your science notebook. List the criteria and constraints.

2 Find a solution.

Your teacher will demonstrate several ways to make sound by blowing. Think about wind instruments you have played or have seen others play.

Next think about how you can use the available materials to make a wind instrument. Will you blow into your instrument or across it? How will your instrument produce different pitches? Do research if needed to help you find your solution.

Sketch a design for your instrument. Share your design with your team. Choose the design that seems most likely to meet the criteria. Have your teacher approve it.

Discuss how you will use your instrument to communicate. Your teacher will give you a list of commands. Come up with a pattern of different pitches to signal each command. Record the meaning of each sound pattern.

A group of musicians in Peru play pan flutes. The pipes of their flutes are different lengths. As a result, they produce sounds with different pitches.

3 Test your solution.

Build your wind instrument. Try it out. Can you make it produce at least three different pitches? If not, adjust it as needed. Make notes on your design to show what you change.

Practice using your instrument to send signals. Test the coded patterns you created for each command on your list. Can you tell the signals apart? If not, change your patterns.

Your teacher will give you a message based on your list of commands. One member of your team will use your instrument to send the message. The rest of your team will listen and decode the message.

Record what happens. Discuss the results of the test. Did your wind instrument and your signal meet the criteria of the problem?

4 Refine or change your solution.

Talk with your team about how you can improve your wind instrument and the way you send signals. Use your ideas to make changes. Then test your revised design. Did your changes make a difference?

Present your wind instrument to the class. Explain how you used different patterns of sound to send a signal. Demonstrate the patterns you used. Then describe the results of your tests. Tell how you used the results of the first test to change your design.

Compare your solution to other teams' solutions. How were their solutions similar to or different from yours? What is one other way you could use your wind instrument to communicate a message? Write about it in your science notebook.

NATIONAL GEOGRAPHIC | Science Career

Animal Tracker

Have you ever seen a group of birds flying together in the same direction and wondered where they were going? Martin Wikelski investigates how and where animals migrate. Sometimes animals travel thousands of miles to meet their basic needs. Wikelski believes that people could learn a lot from studying these patterns of mass movement.

How does Wikelski observe animal movement? He follows them! Wikelski tracks animals either in person or by using communication technology. In one study, he attached tiny radio transmitters to individual sparrows. Using receivers on airplanes, Wikelski and his team tracked the sparrows' movements. They found that the adult migrating birds can find their way even after winds blow them thousands of miles off course.

This radio transmitter is so tiny it can be glued to the back of a tropical orchid bee.

Professor Wikelski places a transmitter on a song sparrow's back. The tracker will send a signal to a satellite that records information about the bird's location.

DCI PS4.C: Information Technologies and Instrumentation. Digitized information can be transmitted over long distances without significant degradation. High-tech devices, such as computers or cell phones, can receive and decode information—convert it from digitized form to voice—and vice versa. (4-PS4-3)
NS Scientific Knowledge Is Based on Empirical Evidence. Science findings are based on recognizing patterns. (4-PS4-1)

NATIONAL GEOGRAPHIC | Explorer

Martin Wikelski is a behavioral ecologist who studies patterns of animal movement. Wikelski has followed birds from Europe to Africa. He has flown with ducks from Sweden to Germany. And he has ridden a motorcycle across the Alps on the trail of songbirds. In addition to birds, he has tracked bats, giant tortoises, bumblebees, and dragonflies.

Citizen Science
Track Bird Life

eBird technology makes it easy to record and send data.

Use a field guide to identify birds. You can make your own field guide, too!

Your class will need to carefully identify and count birds.

DCI PS4.C: Information Technologies and Instrumentation. Digitized information can be transmitted over long distances without significant degradation. High-tech devices, such as computers or cell phones, can receive and decode information—convert it from digitized form to voice—and vice versa. (4-PS4-3)

NS Scientific Knowledge Is Based on Empirical Evidence. Science findings are based on recognizing patterns. (4-PS4-1)

Many people who enjoy watching birds are helping gather data for ornithologists, or scientists who study birds.

What Is Citizen Science?

Citizen scientists are ordinary people who help gather data for real scientific studies. They help professional scientists who could not do all the work on their own. For example, many people who enjoy watching birds are helping gather data for ornithologists, or scientists who study birds.

These citizen scientists are using an online database called eBird to share their observations of birds. A database stores data collected from many sources. The data are organized so that people can easily search and find the information they need. Every month, citizen scientists report millions of bird observations to eBird! All of these observations get added to the database.

Your class will gather data about birds in your area to share with eBird. You'll need to practice identifying your local birds. Field guides can help. They list tips for how to tell birds apart, such as by shape, colors, and calls.

You will work with a partner to observe birds in your area. You will need to follow eBird rules for recording observations, such as when and where they are made. You will use GPS to help accurately identify your location. Then your class can share data to make a class bird list that your teacher can upload to the eBird website.

How will your data be used? Ornithologists use data on eBird to help them track bird movements. Maps on eBird show where different birds live in summer and in winter. eBird data also show which birds might be losing numbers and why. That can help ornithologists take steps to save wild birds.

Wrap It Up!

1. **Describe** What was the most difficult part of the eBird project? What was the most interesting bird you saw?

2. **Explain** How did technology, such as computers and GPS, help you collect and share data?

3. **Analyze** Which were the most common birds your class found?

Check In 📓 My Science Notebook

Congratulations! You have completed *Physical Science*. Let's reflect on what you have learned. Here is a checklist to help you judge your progress. Look through your science notebook to find examples for each statement in the checklist. What could you do better? Write it on a separate page in your science notebook.

▽ Read each item in this list. Ask yourself if you think you did a good job of it.

For each item, select the choice that is true for you: A. Yes B. Not Yet

- I defined and illustrated science vocabulary, science concepts, and main ideas.
- I labeled drawings. I included captions and notes to explain ideas.
- I collected objects, such as photos and magazine or newspaper clippings.
- I used tables, charts, or graphs to record observations and data in investigations.
- I recorded evidence for explanations and conclusions in investigations.
- I described how scientists and engineers answer questions and solve problems.
- I asked new questions.
- I did something else. (Tell about it.)

Reflect on Your Learning

1. Choose one investigation that you found most interesting. Explain why you thought it was most interesting.
2. Choose one main idea that you think was most important to learn about. Explain your thinking.

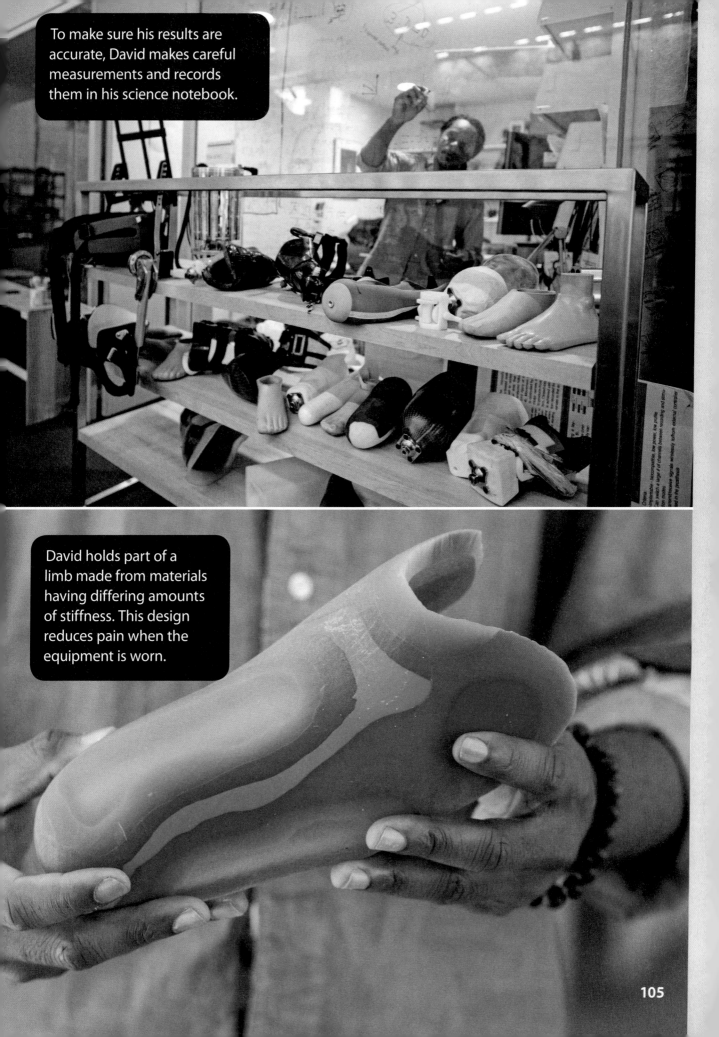

To make sure his results are accurate, David makes careful measurements and records them in his science notebook.

David holds part of a limb made from materials having differing amounts of stiffness. This design reduces pain when the equipment is worn.

NATIONAL GEOGRAPHIC | Explorer

David Moinina Sengeh
Biomedical Engineer
National Geographic Explorer

Let's Explore!

There are many ways of doing science. In *Nature of Science,* you learned that scientists analyze data to develop evidence. They often look for patterns and make inferences to form explanations. I look for patterns that show how the shape of an artificial limb affects its function. I make inferences to improve my designs. As you read, look for ways that scientists analyze data for patterns and make inferences. That includes you, too!

My investigation of the human body relates to life science. Life science is the study of living things and their environments. Here are some questions you might investigate as you read *Life Science*:

- Which parts of a flowering plant produce seeds for new plants to grow?
- How do the internal structures of plants help them reproduce?
- What surprising thing can an elephant do with its feet?
- Why is an animal's brain so important for the animal's survival?

Look at the notebook examples. Do they bring other questions to mind? Write them down. Also, write down your own questions and try to answer them as you read. Let's check in later to review what you have learned!

▼ Collect photos related to main ideas.

▼ Summarize important science ideas in your own words.

Parts of a Flower

Stamens make pollen. Petals can attract insects that carry pollen to a pistil. The pistil develops into a fruit with seeds. Each seed can grow into a new plant.

Animal Senses

I learned that animals use senses to learn about their environment. This helps them survive.

Example: A clouded leopard has sound receptors in its ears. It detects vibrations of a moving ground squirrel. Signals go to the brain. The brain interprets the signals as sound. The leopard can use the information and its memory to catch and eat the ground squirrel!

▶ Connect important science ideas to your world.

My Five Senses

I use my five senses to interact with my environment.

Smell — I use my nose to smell flowers.
Hear — I use my ears to hear birds.
Sight — I use my eyes to see people.
Touch — I use my hands to feel heat.
Taste — I use my tongue to taste apples.

Life Science
Structure, Function, and Information Processing

The Borneo orangutan finds fruit for its next meal.

External Structures of a Wild Rose

Have you ever seen a wild rose plant like the one in the photo? The rose is a type of plant that produces flowers. Its flowers are beautiful, but they are also important to the plant. Like all plants, the wild rose is made up of different kinds of structures. Its external structures are the parts that you can see on the outside of the plant. They include leaves, stems, roots, and flowers.

Leaves, stems, and roots have important functions in the growth and survival of the plant. The wild rose plant also has structures that allow it to reproduce—its flowers. Flowers produce seeds, which can grow into new plants.

Both the color and scent of the wild rose plant's flowers attract insects.

DCI LS1.A: Structure and Function. Plants and animals have both internal and external structures that serve various functions in growth, survival, behavior, and reproduction. (4-LS1-1)
CCC Systems and System Models. A system can be described in terms of its components and their interactions. (4-LS1-1)

Flower The flower of a rose allows the plant to reproduce.

Petal Colorful petals attract bees and other insects, which carry pollen from one flower to another. When an insect leaves pollen on a flower, fruit and seeds can grow.

Stem Stems support the leaves and flowers. As the plant grows, its stems bend toward the light. This behavior helps the leaves get as much sunlight as possible.

Leaf Leaves use water from the soil, carbon dioxide from air, and energy from sunlight to make food for the plant.

Thorn Sharp thorns protect the plant from hungry animals.

Root Roots take in water and dissolved mineral nutrients from the soil. Roots grow downward, allowing them to reach water in the ground.

Wrap It Up! My Science Notebook

1. **List** Name five external structures of a rose plant.

2. **Explain** How do the roots of a rose plant interact with other structures to help the plant grow?

3. **Evaluate** Could a rose plant survive without leaves? Why or why not?

Internal Structures of a Wild Rose

Plants have internal structures that help them grow, survive, and reproduce. These structures exist inside the plant. Many of these structures are hard to see without a magnifying glass or microscope.

The many stamens of this wild rose flower surround the pistil in the center.

Flower In the center of the flower are the stamens and pistil. **Stamens** make pollen. For fruit to develop, pollen must be transferred to the pistil. Then the **pistil** develops into a fruit with seeds inside. Each seed can grow into a new plant.

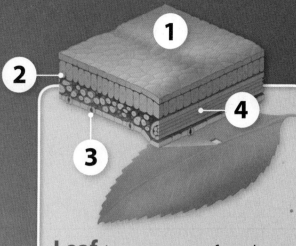

Leaf Leaves use water from the soil, carbon dioxide from the air, and energy from sunlight to make food for the plant. Leaves are made up of several parts.

1. The outer layer protects the leaf and keeps it from drying out.
2. In the middle is the food-making layer.
3. Openings in the bottom of the leaf let air into the food-making layer.
4. Veins are made up of tiny tubes. Some tubes carry water to the leaf. Other tubes carry food from the leaves to the rest of the plant.

Stem Inside each stem are bundles of tiny tubes. Some tubes carry water from the roots up to the leaves and flowers. Other tubes carry food from the leaves to the rest of the plant.

Roots Tiny hairs on the roots take in water and mineral nutrients from the soil.

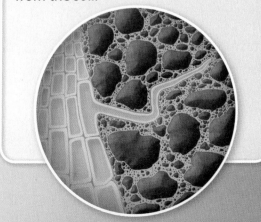

Wrap It Up!

1. **Identify** What are the structures of a leaf? What are their functions?

2. **Cause and Effect** How do the stamens of a flower help a plant reproduce?

3. **Analyze** How is a plant's stem important to the function of the plant's other parts?

NATIONAL GEOGRAPHIC | Think Like a Scientist

Construct an Argument

A buttercup plant and a wild rose look very different. But a buttercup, like a wild rose, has external and internal structures that help it survive.

Use sticky notes to label the different external and internal structures of the buttercup plant. Then follow the process below.

1. **List.** *My Science Notebook*
   What external structures did you label? What internal structures did you label?

2. **Compare.**
   Work with a group. Compare your labels. Work together until everyone in your group has all the buttercup plant's structures labeled. Then compare the buttercup plant's structures to the wild rose's structures in the two previous lessons.

3. **Construct an argument.**
   Have each person in the group choose one labeled *external structure* and one labeled *internal structure*. Write an explanation arguing how, as with the wild rose, these structures help the buttercup plant grow, survive, or reproduce. Use evidence from the diagram to support your argument. Also use evidence from lessons on the internal and external structures of a wild rose.

4. **Generalize.**
   Come back together as a group. Present your arguments for the structures you labeled. Together, write a summary explaining how the structures of the buttercup plant and the wild rose help them survive.

115

External Structures of an Elephant

What animal uses its nose to put food in its mouth and its feet to sense sound? An elephant! An elephant's body is made up of many different structures that allow it to grow, survive, and respond to its surroundings. The photo shows some of the external structures of an Asian elephant.

The Asian elephant is the biggest animal in Asia. To get enough energy to survive, an elephant needs to eat an enormous amount of food. No wonder elephants spend most of their time looking for food. They eat grasses, leaves, roots, bark, and fruit.

Elephants use a wide variety of sounds to communicate—trumpeting, roaring, snorting, grunting, and barking. They also make rumbling sounds that are too low for humans to hear. These low sounds can travel through the ground for long distances—as far as 32 kilometers (20 miles). Elephants use their ears and feet to sense vibrations from sounds traveling through air and underground.

> **Skin** Tough, wrinkled skin protects the elephant's internal organs. The skin also keeps the elephant cool. To protect its skin from getting too much sun, an elephant may roll in mud or cover itself with dust.

DCI LS1.A: Structure and Function. Plants and animals have both internal and external structures that serve various functions in growth, survival, behavior, and reproduction. (4-LS1-1)
CCC Systems and System Models. A system can be described in terms of its components and their interactions. (4-LS1-1)

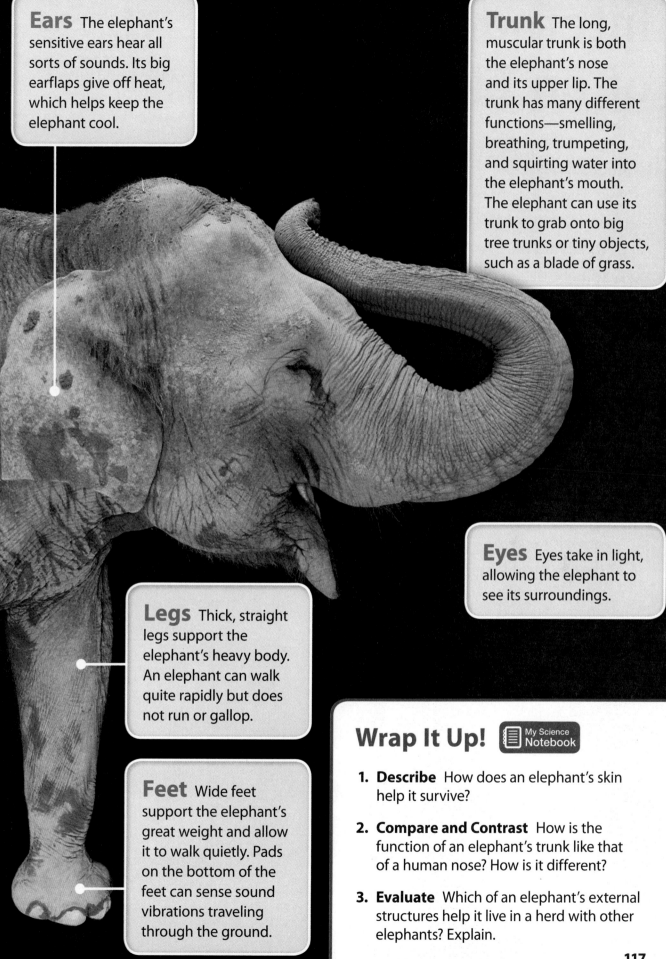

Ears The elephant's sensitive ears hear all sorts of sounds. Its big earflaps give off heat, which helps keep the elephant cool.

Trunk The long, muscular trunk is both the elephant's nose and its upper lip. The trunk has many different functions—smelling, breathing, trumpeting, and squirting water into the elephant's mouth. The elephant can use its trunk to grab onto big tree trunks or tiny objects, such as a blade of grass.

Eyes Eyes take in light, allowing the elephant to see its surroundings.

Legs Thick, straight legs support the elephant's heavy body. An elephant can walk quite rapidly but does not run or gallop.

Feet Wide feet support the elephant's great weight and allow it to walk quietly. Pads on the bottom of the feet can sense sound vibrations traveling through the ground.

Wrap It Up! My Science Notebook

1. **Describe** How does an elephant's skin help it survive?

2. **Compare and Contrast** How is the function of an elephant's trunk like that of a human nose? How is it different?

3. **Evaluate** Which of an elephant's external structures help it live in a herd with other elephants? Explain.

Internal Organs of an Elephant

An elephant's internal organs serve various functions. They allow the elephant to grow and survive. These functions include providing the elephant's body with food, water, and oxygen. All of the living parts of the elephant's body require these materials in order to survive.

Lungs The lungs take in oxygen from the air and release carbon dioxide. Blood traveling through the lungs picks up oxygen.

Stomach The large stomach stores food and begins the process of breaking it down. Food then travels to the intestine.

Intestines Most of an elephant's food is digested in the small intestine. Bacteria at the end of the small intestine help break down the food. Sugars and other chemicals from food are taken up by blood in the intestine walls. Undigested food moves from the small intestine to the large intestine.

Liver The liver produces many chemicals that are necessary for the functions of an elephant's body. For example, the liver produces bile. Bile helps break down fats during the process of digestion.

DCI LS1.A: Structure and Function. Plants and animals have both internal and external structures that serve various functions in growth, survival, behavior, and reproduction. (4-LS1-1)
CCC Systems and System Models. A system can be described in terms of its components and their interactions. (4-LS1-1)

Esophagus The esophagus is the tube that carries food from the elephant's mouth to its stomach.

Brain The large and highly developed brain makes the elephant very intelligent. The brain controls all of the functions of the elephant's body. It processes information, coordinates the elephant's behavior, and allows it to respond to its surroundings.

Teeth Large, flat teeth grind up food, starting the process of digestion.

Heart The large, muscular heart pumps blood throughout the elephant's body. Blood carries food and oxygen to all parts of the elephant's body.

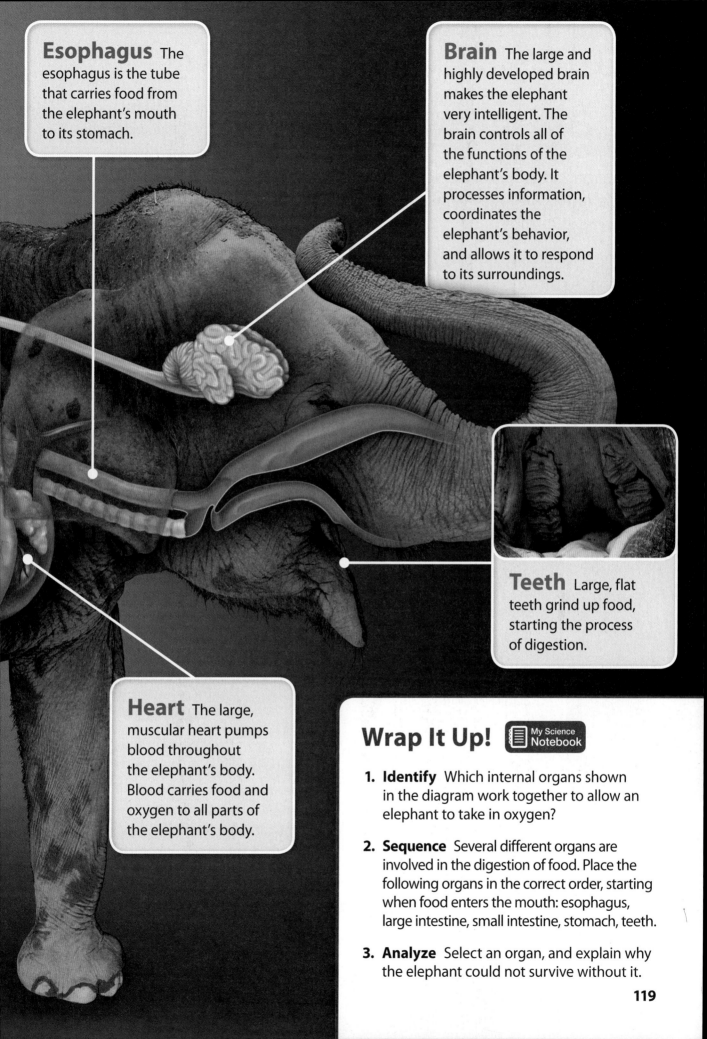

Wrap It Up! My Science Notebook

1. **Identify** Which internal organs shown in the diagram work together to allow an elephant to take in oxygen?

2. **Sequence** Several different organs are involved in the digestion of food. Place the following organs in the correct order, starting when food enters the mouth: esophagus, large intestine, small intestine, stomach, teeth.

3. **Analyze** Select an organ, and explain why the elephant could not survive without it.

119

Bones and Muscles of an Elephant

The internal structures of an elephant include its bones and muscles. Bones support the elephant's body and protect its internal organs. All of the bones in an elephant make up its skeleton. Muscles are attached to bones and work with the bones to move parts of the elephant's body.

Ribs The ribs protect the heart and lungs.

Structure of a bone The outside of a long leg bone is hard and compact. The inside of the bone is spongy. Blood vessels in the spongy part of the bone bring nutrients and oxygen to the bone.

Skeletal muscles Skeletal muscles are attached to bones. When a skeletal muscle contracts, it pulls on the bone and makes it move.

Pelvis The bones of the pelvis are beneath the muscles. The pelvis provides a frame that supports the back legs. Joints in the pelvis allow the elephant to move its legs so it can walk or swim.

DCI LS1.A: Structure and Function. Plants and animals have both internal and external structures that serve various functions in growth, survival, behavior, and reproduction. (4-LS1-1)
CCC Systems and System Models. A system can be described in terms of its components and their interactions. (4-LS1-1)

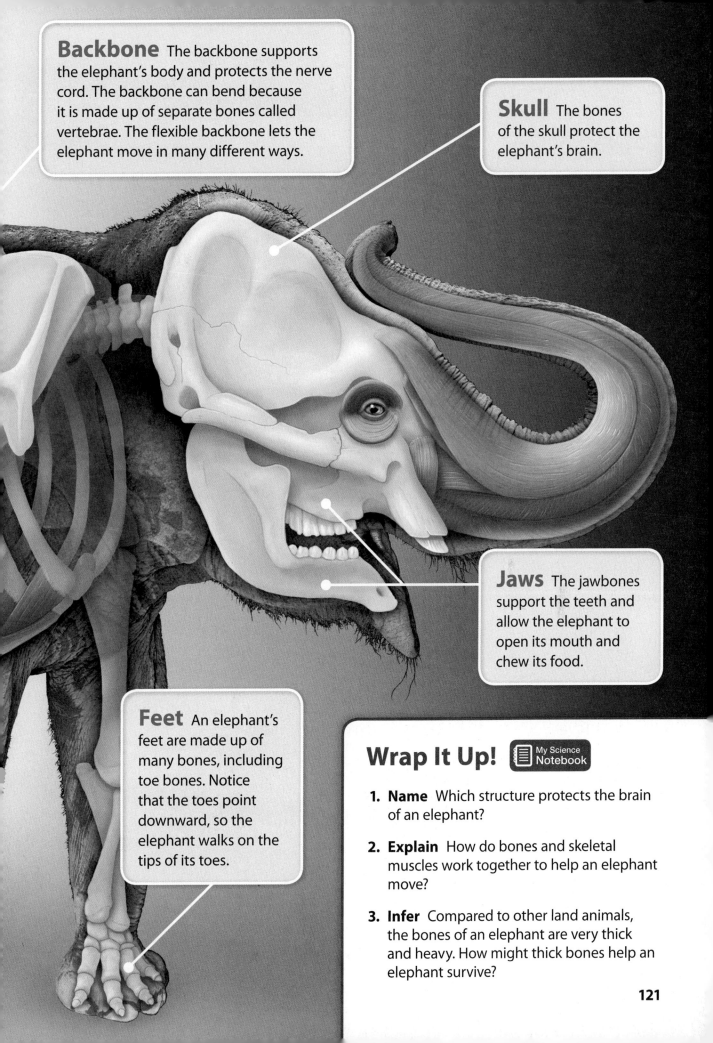

Backbone The backbone supports the elephant's body and protects the nerve cord. The backbone can bend because it is made up of separate bones called vertebrae. The flexible backbone lets the elephant move in many different ways.

Skull The bones of the skull protect the elephant's brain.

Jaws The jawbones support the teeth and allow the elephant to open its mouth and chew its food.

Feet An elephant's feet are made up of many bones, including toe bones. Notice that the toes point downward, so the elephant walks on the tips of its toes.

Wrap It Up! My Science Notebook

1. **Name** Which structure protects the brain of an elephant?

2. **Explain** How do bones and skeletal muscles work together to help an elephant move?

3. **Infer** Compared to other land animals, the bones of an elephant are very thick and heavy. How might thick bones help an elephant survive?

Stories in Science

Listening to Elephants

A scientist works to place recording equipment high in a tree, where it will not bother the elephants.

Once this recording equipment is in place, it can record sound on its own for up to six months.

Members of an elephant herd use infrasound to keep in touch if they are far apart. The low sounds begin as rumbles in their throats.

DCI LS1.A: Structure and Function. Plants and animals have both internal and external structures that serve various functions in growth, survival, behavior, and reproduction. (4-LS1-1)
DCI LS1.D: Information Processing. Different sense receptors are specialized for particular kinds of information, which may be then processed by the animal's brain. Animals are able to use their perceptions and memories to guide their actions. (4-LS1-2)
NS Scientific Knowledge Is Based on Empirical Evidence. Science findings are based on recognizing patterns. (4-PS4-1)

Elephants have a secret language...
Katy Payne is listening in.

One day in 1984, Katy Payne was at the Portland Zoo. She was near a concrete wall that separated two elephants. Katy didn't hear anything odd. But she did *feel* something odd: a rumbling in her body. She suspected it was a form of infrasound. Infrasound is sound that is too low for humans to hear, but it can be felt as vibrations. Katy wondered about the vibrations she felt. Were the elephants causing them?

Katy's life experiences uniquely prepared her to answer this question. Growing up, Katy spent lots of time outdoors. Then she studied music and biology at college. After college, she combined these two interests and became an acoustic biologist. She studied the science of sound.

Katy and her husband had already studied whale sounds. They discovered that whales sing! Whale songs contain different patterns of melodies, rhythms, and rhymes. Whales use the songs to communicate.

Katy went back to the zoo to see the elephants. This time she brought equipment that could record the low rumbling sounds. Katy was right! Elephants were using infrasound to communicate in some way.

Katy headed to Africa to study wild elephants. She began recording their rumbles as part of the Elephant Listening Project. Katy and her team have since collected thousands of hours of recordings.

Katy learned that elephants use infrasound to keep in touch when they are separated. Elephant rumbles cause sound vibrations that travel long distances through the ground. Elephants sense the vibrations with their feet. Family members can even recognize each other's rumbles!

Wrap It Up!

1. **Describe** How do elephants use infrasound to communicate?

2. **Explain** How do you think Katy's experience with music helps her work as an acoustic biologist?

3. **Infer** How might an elephant tell family members' rumbles apart?

NATIONAL GEOGRAPHIC | Think Like a Scientist

Construct an Argument

A wolf and an elephant look very different. But a wolf, like an elephant, has external and internal structures that help it survive.

Use sticky notes to label the different internal and external structures of the wolf. Then follow the process below.

1. **List.**
   What external structures did you label? What internal structures did you label?

2. **Compare.**
   Work with a group. Compare your labels. Did you all label the same structures? Work together until everyone in your group has all the wolf's structures labeled. Then compare the wolf's structures to the elephant's structures in the previous lessons.

3. **Construct an argument.**
   Have each person in the group choose one labeled *external structure* and one labeled *internal structure*. Write an explanation arguing how, as with an elephant, these structures help the animal grow, survive, behave in certain ways, or reproduce. Use evidence from the diagram to support your explanation. Also use evidence from the lessons on an elephant's internal and external structures.

4. **Generalize.**
   Come back together as a group. Present your arguments for the structures you labeled. Together, write a summary explaining how the structures of the wolf and the elephant help them survive.

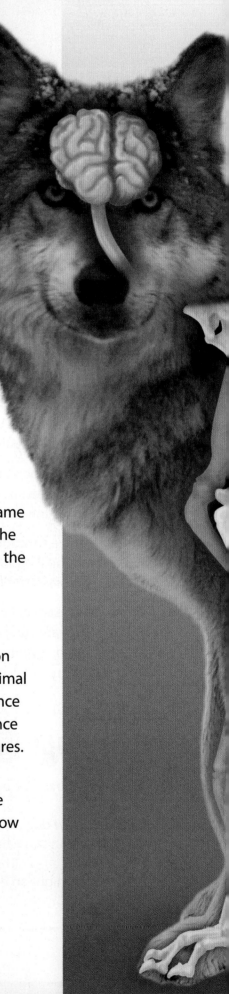

PE 4-LS1-1. Construct an argument that plants and animals have internal and external structures that function to support survival, growth, behavior, and reproduction.

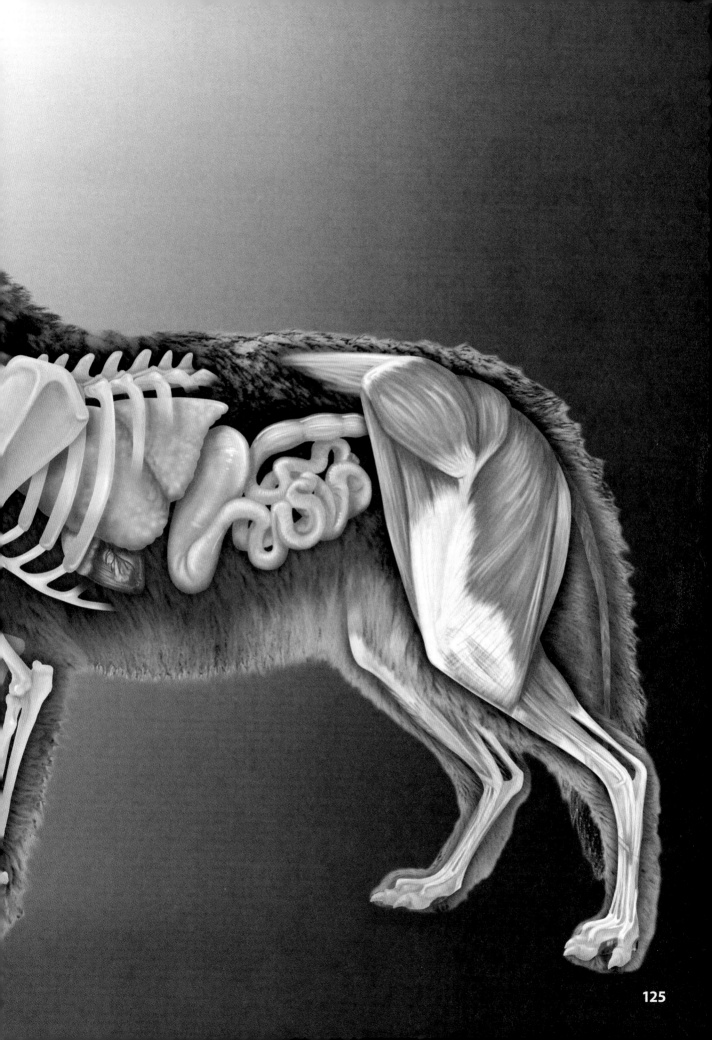

Animal Senses

The clouded leopard is a fierce cat that lives in the forests of Southeast Asia. It uses its senses to learn about its surroundings. Each sense receptor responds to a particular kind of information and sends signals to the brain. The brain processes those signals so they have meaning for the cat.

The clouded leopard uses its keen sense of hearing to learn when its predators and prey are nearby. For example, the sounds of a ground squirrel's movements travel through the air as vibrations. When those vibrations enter the cat's ears, sound receptors send signals to the cat's brain. The brain interprets those signals as the noises made by a ground squirrel. The cat uses that perception and its memories of hunting other animals to catch the ground squirrel. Read the captions to find out how the clouded leopard's other senses aid it in processing information.

DCI LS1.D: Information Processing. Different sense receptors are specialized for particular kinds of information, which may be then processed by the animal's brain. Animals are able to use their perceptions and memories to guide their actions. (4-LS1-2)
CCC Systems and System Models. A system can be described in terms of its components and their interactions. (4-LS1-2)

Hearing Sound receptors in the ear respond to vibrations in the air, causing signals to travel to the brain. The brain perceives them as different sounds.

Sight Light receptors in the clouded leopard's eyes respond to light and send signals to the brain. The brain then processes those signals, letting the clouded leopard know what it is seeing.

Smell Smell receptors in the cat's nose are sensitive to chemicals in the air. Those receptors send signals to the brain, which interprets them as different odors.

Touch When the clouded leopard's whiskers brush against an object, touch receptors send signals to its brain, which processes them. This lets the cat know that its whiskers are touching something.

Taste Taste buds on the clouded leopard's tongue respond to chemicals in food, sending signals to the brain. The brain interprets those signals as different flavors.

Wrap It Up!

1. **Identify** What senses does a clouded leopard use to know what is in its environment?

2. **Relate** How is a clouded leopard's brain related to its senses?

Light and Sight

How does a clouded leopard see a ground squirrel? First, sunlight **reflects** from, or bounces off, the squirrel. The reflected light travels through the air. When the light enters the cat's eye, it hits light receptors at the back of the eyeball. Those receptors send signals to the brain. The brain processes the signals, so the clouded leopard understands that it is seeing a ground squirrel.

DCI PS4.B: **Electromagnetic Radiation.** An object can be seen when light reflected from its surface enters the eyes. (4-PS4-2)

DCI LS1.D: **Information Processing.** Different sense receptors are specialized for particular kinds of information, which may be then processed by the animal's brain. Animals are able to use their perceptions and memories to guide their actions. (4-LS1-2)

Trace the path of light from the sun to the leopard's eyes. The leopard's brain interprets signals from its eyes. Then the cat can pounce on its prey.

Clouded leopards have good eyesight. Their forward-facing eyes allow them to judge distances as they climb trees and hunt prey.

Wrap It Up!

1. **Define** What does the word *reflect* mean?
2. **Explain** How does light from the sun allow an animal to see?
3. **Infer** Clouded leopards often hunt at night. How do you think they are able to see objects at night?

Investigate

How We See

? How can you model the idea that light allows objects to be seen?

Most objects do not give off their own light. You can only see such objects when light from another source bounces off of them and enters your eyes. Buildings are full of lamps and other light sources that enable us to see objects in spaces that sunlight does not reach. In this activity, you will explore how you see objects using a light source.

Materials

flashlight	classroom objects	shoebox

1. Your teacher will give your group a shoebox with an object inside it. Then your teacher will turn out the lights in the room.

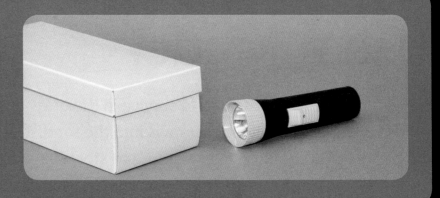

2. Open the shoebox. Shine a flashlight on the object in the box. Look at the object in the box. Then turn on the lights in the room.

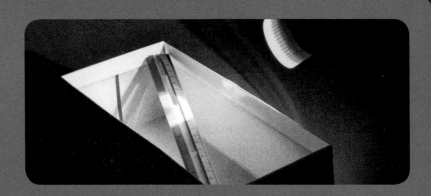

3. Draw a model that shows how light reflecting from the object and entering your eye allows you to see the object. Write captions that explain how light allows the object to be seen.

4. Explain how you could revise your model so that you do not need to open the lid of the box to view the objects.

Explore on Your Own

How could you use a flashlight and a mirror to reflect light in order to view an object in the dark? Plan and carry out your own investigation. Record your observations. Compare the results of your investigations.

Wrap It Up!

1. **Explain** Use information from your model to explain how light from the flashlight reached your eye.

2. **Apply** Is it possible to see an object when there is no light? Why or why not? What about when you close your eyes?

NATIONAL GEOGRAPHIC | Think Like a Scientist

Use a Model

A beach mouse uses its senses to survive in the grassy sand dunes of Florida. It digs burrows into the sand. It uses the burrows to keep safe, raise young, and store seeds for food. The mice look for food mainly at night. The eastern diamondback rattlesnake preys on beach mice and other small animals. Imagine that this mouse has just come out of its burrow. It is late at night, and it is very dark. Suddenly, an eastern diamondback rattlesnake notices the mouse!

1. **Make a model.**
   Design a model that shows how the mouse and the snake might receive information through several of their senses as they search for food. Include how the mouse and snake might process and respond to the information. Think about how some information might be stored as memories.

2. **Discuss your model.**
   Work with a partner. Use your models to talk about how the mouse and snake might receive, process, and react to information.

3. **Revise your model.**
   Do research to find out how the mouse and snake gather information from the environment. Then revise your model to show what information the mouse and the snake receive and process.

4. **Share your model.**
   Meet with another pair of students. Discuss how the models you made are the same and different. Take turns using your model to explain how the snake and the mouse shown in this scene would respond based on information from their senses.

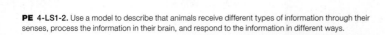

The beach mouse spends little time outside its burrow during daylight hours.

The eastern diamondback rattlesnake preys on rats, mice, squirrels, and birds.

Animal Super Senses

Scientists have discovered that some animals have "super senses." These animals are especially good at seeing, smelling, hearing, touching, or tasting.

A chameleon is one animal with a super sense of sight. Its eyes can move and focus in different directions at the same time. In this way, it is able to see in nearly every direction at once. When a chameleon spots prey, it can respond by shooting out its long, sticky tongue to catch it. When a chameleon spots danger, it can quickly run to escape.

Now it is your turn to explore super senses. Work with a partner to identify two animals that have special ways of sensing the world around them. Research and write about how each animal uses its super sense to receive, process, and respond to information.

DCI LS1.A: Structure and Function. Plants and animals have both internal and external structures that serve various functions in growth, survival, behavior, and reproduction. (4-LS1-1)
DCI LS1.D: Information Processing. Different sense receptors are specialized for particular kinds of information, which may be then processed by the animal's brain. Animals are able to use their perceptions and memories to guide their actions. (4-LS1-2)
SEP Obtaining, Evaluating, and Communicating Information. Obtain and combine information from books and other reliable media to explain phenomena. (4-ESS3-1)

The Challenge

Your challenge is to identify and research two animals that have super senses. You will make a booklet, poster, or computer slide presentation and share it with your class.

A chameleon's eyes are located on opposite sides of its head. How might this help it get a better view?

STEM RESEARCH PROJECT (continued)

1 Select a topic.

Work with your partner to brainstorm a list of animals you might research more about. Include any interesting animals you have learned about from books, TV, movies, or websites. Maybe you have even seen an animal use one of its senses in an unusual way.

Write down any ideas that come to you in your science notebook. If you have trouble coming up with animals to research, ask your teacher for ideas.

Choose two animals that you would like to research. Have your teacher approve them. What are some questions you want to research about these animals and their senses? Record your questions in your science notebook. Identify key words in your questions. You can use the key words to help guide your research.

2 Plan and conduct research.

Make a plan with your partner to do research using online and printed sources. Try to answer your questions from Step 1. Your research should also focus on these questions:

- What "super sense" does the animal use, and what structures are involved?
- How does the animal use its sense to receive, process, and respond to information? Does it respond right away or store memories?
- How does this animal's sensing ability compare to that of other animals, including humans?
- What is one more interesting fact about the animal?

Be sure to choose sources that have accurate information. Find at least two sources to support each fact. Record the information you find, including the source. Use outlining or graphic organizers to organize your notes.

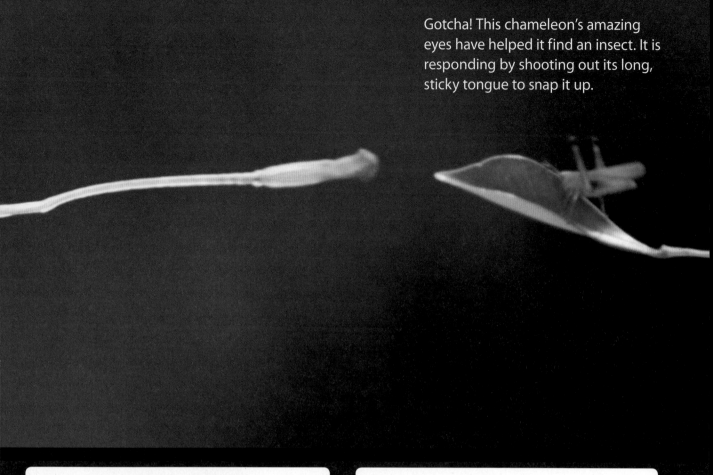

Gotcha! This chameleon's amazing eyes have helped it find an insect. It is responding by shooting out its long, sticky tongue to snap it up.

3 Draft and finalize your report.

Your report will be in the form of a booklet, poster, or computer slide presentation. In your report, you should summarize the main ideas and the most important details of your research. The information you present should be in your own words.

Look back at your questions in Steps 1 and 2 to be sure you are including all the information you need. Organize the information in a way that is easy to follow. For example, you might divide the information into sections. Each section could explain one of the answers you found in your research. Include at least one picture of each animal, too.

Review and rewrite the draft of your report to make it the best it can be. Do more research to add information as needed. Make the final draft of your booklet, poster, or slide presentation.

4 Present your report.

You will present your report to the class. Work with your partner to decide who will give each part of the presentation. Decide how you will display pictures or other visual information.

Practice giving your part of the report aloud. Your oral presentation should express main ideas that are supported by important details. Ask your partner to give you feedback to help you improve your presentation.

With your partner, present your report to the class. Put information in a logical order. Use descriptions, facts, and details to describe the animals and their unique senses. Remember to give an additional interesting fact about each animal. Speak loudly and slowly. Answer any questions your classmates may have.

Listen as your classmates present their reports. How many different animal super senses did your classmates identify and report on?

NATIONAL GEOGRAPHIC | Science Career

Dog Whisperer

What can you do if your dog won't behave? Maybe it barks all the time or chews up shoes. Maybe it fights with other dogs or jumps up on strangers.

Cesar Millan knows how to fix all these problems. How? By using dog psychology. Psychology is the science of the mind and behavior. Cesar observes how dogs use their senses to respond to their environment. He studies how these responses are stored as memories that guide their behavior.

Cesar uses his understanding of dogs to correct the behavior of pet dogs that are out of control. But mostly, Cesar shows dog owners how to change the way they treat their pets. Here's the surprising fact: When owners correct the way they treat their pets, their pets almost always stop misbehaving!

Cesar is called the "dog whisperer" because he has a special talent for interacting with dogs in ways that improve their behavior.

DCI LS1.D: Information Processing. Different sense receptors are specialized for particular kinds of information, which may be then processed by the animal's brain. Animals are able to use their perceptions and memories to guide their actions. (4-LS1-2).
NS Scientific Knowledge Is Based on Empirical Evidence. Science findings are based on recognizing patterns. (4-PS4-1)

Cesar Millan is a dog trainer, author, and star of the National Geographic Channel program "The Dog Whisperer." Originally from Mexico, Cesar now lives in the United States. Cesar founded and runs the Dog Psychology Center in Los Angeles, where he rehabilitates dogs with severe behavior problems. Cesar has won many awards, including one from the Humane Society for his work with dogs from shelters.

Science Career (continued)

NGL Science How do you rehabilitate dogs?

Cesar Millan I don't train dogs to respond to commands such as "sit" or "stay." Instead I try to connect with the dog's mind and how it naturally responds to its environment to help correct unwanted behaviors.

NGL Science How did you learn about dogs?

Cesar Millan As a child, I spent a lot of time on my grandfather's farm in Mexico. The working dogs on the farm were my true teachers in the art and science of dog psychology. I loved to watch dogs play with one another. The more hours I spent watching them, the more questions came into my mind. How did they coordinate their activities? How did they communicate with one another?

NGL Science How did you develop your methods of training dogs?

Cesar Millan My way of training came directly from my observations of dogs on my grandfather's farm. I interact with the dogs the way they interact with one another.

NGL Science Do you also work with the dog owners?

Cesar Millan The dog owner often thinks that the problem lies with the dog. But the problem is usually with the way the owner treats his dog. I often say, "I rehabilitate dogs, but I train people." My formula is simple: For a balanced, healthy dog, a human must share exercise, discipline, and affection, in that order!

Cesar plays with this dog as if he were a dog himself.

Check In 📓 My Science Notebook

Congratulations! You have completed *Life Science*. Let's reflect on what you have learned. Here is a checklist to help you judge your progress. Look through your science notebook to find examples for each statement in the checklist. What could you do better? Write it on a separate page in your science notebook.

▼ Read each item in this list. Ask yourself if you think you did a good job of it.

For each item, select the choice that is true for you: A. Yes   B. Not Yet

- I defined and illustrated science vocabulary, science concepts, and main ideas.
- I labeled drawings. I included captions and notes to explain ideas.
- I collected objects, such as photos and magazine or newspaper clippings.
- I used tables, charts, or graphs to record observations and data in investigations.
- I recorded evidence for explanations and conclusions in investigations.
- I described how scientists and engineers answer questions and solve problems.
- I asked new questions.
- I did something else. (Tell about it.)

Reflect on Your Learning

1. Choose one science idea that you think was most important to learn about. Explain your thinking.
2. What is one way your understanding of a science idea changed? Describe how your thinking has changed.

This artificial limb is custom designed for a single person to wear. David studied a model of the person's own body to be sure the limb would fit perfectly.

143

Explorer

David Moinina Sengeh
Biomedical Engineer
National Geographic Explorer

Let's Explore!

In *Nature of Science,* you learned that creativity is an important part of science. It helps scientists see things in new ways. I once used creativity to solve the problem of providing electricity to people living far from cities. I found a way to use tiny organisms living in soil to help produce electricity! As you read, look for ways scientists think creatively. That includes you, too!

My investigation of soil life relates to Earth science. Earth science is the study of Earth's land, water, and atmosphere. Here are some questions you might investigate as you read *Earth Science*:

- What are the different natural regions across the United States, and what types of plants and animals live in each region?
- How can wind, water, or a tree change the shape of a huge rock?
- What are some ways people can reduce the effects of a dangerous earthquake?
- How do fossils give clues about the land in the past?

Look at the notebook examples for ideas of other questions to ask. Let's connect again later to review what you have learned!

▼ Connect science ideas to real-world examples.

Changing Earth's Surface

weathering

deposition

natural hazards

living things

I have seen how living things change the land. My yard has holes where a chipmunk has dug holes for its burrow.

▼ Reflect on what you have learned.

Important Things I Learned

The amount of rainfall in a region affects the types of plants and animals found there. Rainfall also helps to shape the land.

Earthquakes, tsunamis, and volcanoes are examples of natural hazards.

The Mariana Trench is the deepest place in the world's oceans.

Engineers use building designs and materials to reduce damage caused by earthquakes.

▶ Analyze data to look for patterns.

Earthquakes and Volcanoes

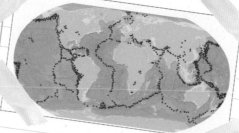

Describe the pattern in the locations of the active volcanoes.

Describe the pattern in the location of earthquake activity.

Earth Science

Earth's Systems: Processes that Shape the Earth

The Sakurajima volcano in southern Japan has eruptred almost continuously since 1955.

The amount of rain that falls during the year varies from region to region. The map shows areas that receive different amounts of rain. Rainfall helps to shape the land and affects the types of plants and animals found in a region.

Pacific Northwest Forest

The Pacific Northwest coast is in a very rainy region. Some places within that region receive more than 250 centimeters (100 inches) of rain each year! This wet weather supports temperate rain forests with tall trees and many different kinds of plants and animals.

Southwest Desert

The deserts of the southwest receive less than 30 centimeters (12 inches) of rain a year. The living things that live here can survive with little water.

Central Plains Grassland

The middle part of the United States receives moderate rain—about 51 centimeters (20 inches) each year. This much rainfall supports grasslands. It provides enough water for grasses, but not enough for most trees.

Eastern Temperate Forest

The eastern United States receives an average of around 127 centimeters (50 inches) of rain throughout the year. The plentiful rainfall supports temperate forests with many plants and animals.

Average Annual Precipitation
centimeters (inches)
- Over 180 (70)
- 100–180 (40–70)
- 50–100 (20–49)
- 0–50 (0–20)

Wrap It Up!

1. **Define** About how much rainfall does the southwest desert region receive?

2. **Contrast** How is a grassland different from a temperate forest? Contrast the amount of rain and the kinds of plants in the two places.

3. **Cause and Effect** How does the amount of rainfall affect the living things that can live in a region?

149

Pacific Northwest Forest

If you visit the Pacific Northwest, you'll see many tall trees. In fact, this region is home to the world's tallest trees! The rainy weather and mild temperatures support the growth of lush forests. Most of the trees are evergreen, which means that they keep their needle-shaped leaves all year long.

Many flowering shrubs and trees, such as currants and dogwoods, also live in these forests. Wildflowers, mosses, and ferns grow in the moist soil. This forest is home to many different kinds of animals, from insects and amphibians to birds and mammals.

When it is not raining in the Pacific Northwest, the trees are still often wrapped in moisture in the form of fog.

DCI ESS2.A: Earth Materials and Systems. Rainfall helps to shape the land and affects the types of living things found in a region. Water, ice, wind, living organisms, and gravity break rocks, soils, and sediments into smaller particles and move them around. (4-ESS2-1)
CCC Cause and Effect. Cause and effect relationships are routinely identified, tested, and used to explain change. (4-ESS2-1)

The sideband snail slides over the wet ground and decaying leaves on the forest floor.

The colorful western tanager eats insects, fruits, and seeds. This male tanager is sitting in a spruce tree in an Oregon forest.

Many wildflowers, such as these iris, grow in the damp soil.

Many long growing seasons with plentiful rain have allowed this Douglas fir tree to grow very tall.

Wrap It Up!

1. **Describe** What is the weather like in the Pacific Northwest?

2. **Identify** What kinds of trees are most common in the Pacific Northwest?

3. **Cause and Effect** How might this area of the Pacific Northwest look different if it received very little rain?

Southwest Desert

The Sonoran Desert in the southwestern United States looks very different from that of the Pacific Northwest. Why? This region receives very little rain. As a result, trees that grow in mild, rainy regions cannot grow here. Instead, plants that can survive with little water live in this dry region. Desert plants grow farther apart from each other than those in a forest, too. The animals that live in the desert must also be able to survive in harsh conditions without much water.

The tough, scaly skin of reptiles, such as this iguana, slows the loss of water from their bodies. This allows them to survive in the desert.

The saguaro cactus provides shelter around the nest of these Harris's hawks.

DCI ESS2.A: Earth Materials and Systems. Rainfall helps to shape the land and affects the types of living things found in a region. Water, ice, wind, living organisms, and gravity break rocks, soils, and sediments into smaller particles and move them around. (4-ESS2-1)
CCC Cause and Effect. Cause and effect relationships are routinely identified, tested, and used to explain change. (4-ESS2-1)

Cacti such as the tall, forked saguaro grow in the Sonoran Desert.

Plants that live in deserts must have ways to save water. The barrel cactus stores water in its thick stem.

Some plants, such as this purple mat, survive the dry weather as seeds. When there is enough rain, these plants quickly sprout, grow, flower, and set seeds.

Wrap It Up! My Science Notebook

1. **Recall** What characteristics do animals have that help them survive in the Sonoran Desert?

2. **Explain** How do the characteristics of a barrel cactus help this plant to survive in the Sonoran Desert?

3. **Infer** Most amphibians have moist skin and lay their eggs in water. Would you expect there to be many kinds of amphibians in the Sonoran Desert? Explain.

Central Plains Grassland

The central portion of the United States receives more rain than the desert but less rain than the forest. The region supports grasslands, also called prairies. This large area of land is covered in grasses and many kinds of wildflowers, but few trees.

Bison adapt to the changing seasons by shedding their thick winter coat when the weather becomes warm.

This prairie chicken feeds on the grasses growing in a Nebraska prairie.

DCI ESS2.A: Earth Materials and Systems. Rainfall helps to shape the land and affects the types of living things found in a region. Water, ice, wind, living organisms, and gravity break rocks, soils, and sediments into smaller particles and move them around. (4-ESS2-1)
CCC Cause and Effect. Cause and effect relationships are routinely identified, tested, and used to explain change. (4-ESS2-1)

The weather in the grasslands varies from season to season. Winters are cold and summers are hot and dry. Prairie plants grow long roots that can absorb and store water. During extremely dry periods, fires are common. Many prairie plants regrow after a fire because their roots are protected below the ground.

Few tree species grow here because there is not enough rainfall and fires destroy saplings.

These colorful flowers attract bees and wasps that help the plant reproduce.

Prairie grasses have adaptations that help them survive the long, hot summers.

Wrap It Up! My Science Notebook

1. **Describe** What is the weather like in the prairie during the summer?

2. **Cause and Effect** Why aren't most prairie plants killed by fires?

3. **Infer** When do the grasslands get most of their rainfall?

Eastern Temperate Forest

The mid-Atlantic region receives rain and snow throughout the year that supports temperate forests. Temperate forests are wet environments. These forests are similar to those found in the Pacific Northwest, with one major difference. Most trees in temperate forests are deciduous, which means they lose their leaves in fall and produce new ones in the spring.

Abundant rainfall in temperate forests encourages the growth of many trees, shrubs, and other plants.

DCI ESS2.A: **Earth Materials and Systems.** Rainfall helps to shape the land and affects the types of living things found in a region. Water, ice, wind, living organisms, and gravity break rocks, soils, and sediments into smaller particles and move them around. (4-ESS2-1)
CCC **Cause and Effect.** Cause and effect relationships are routinely identified, tested, and used to explain change. (4-ESS2-1)

Deciduous trees drop their leaves before winter to help preserve energy when it is cold. During warmer months, large leaves help low-growing forest plants absorb sunlight.

White-tailed deer thrive in temperate forests.

The large leaves of the trillium help it absorb the small amount of sunlight that reaches the forest floor.

The Northern Saw-whet Owl makes its home in the eastern temperate forest. Owls live in the tall trees and hunt mice, chipmunks, and other small animals.

Flowering trees, such as this dogwood, bloom in the eastern temperate forest in spring.

Wrap It Up!

1. **Define** What are deciduous trees?
2. **Compare and Contrast** How do temperate forests differ from forests found in the Pacific Northwest?
3. **Infer** What characteristics of the temperate forest make it a good home for many animals?
4. **Cause and Effect** Why do many trees in temperate forests appear to be dead in winter?

Weathering

Landforms such as mountains, valleys, and plains are shaped and changed over millions of years. These changes to Earth's surface include three processes. The first process is weathering. **Weathering** happens when rocks break apart, wear away, or dissolve into smaller particles, or when the materials in the rock are changed. The smaller particles are called **sediment.** Wind, water, ice, chemicals, and even plants can cause weathering.

DCI ESS2.A: Earth Materials and Systems. Rainfall helps to shape the land and affects the types of living things found in a region. Water, ice, wind, living organisms, and gravity break rocks, soils, and sediments into smaller particles and move them around. (4-ESS2-1)
CCC Cause and Effect. Cause and effect relationships are routinely identified, tested, and used to explain change. (4-ESS2-1)

Weathering by water and wind helped shape this rock into its arched shape. Eventually more weathering could weaken it until gravity causes it to fall.

Wrap It Up!

1. **Recall** What can cause weathering?

2. **Explain** Describe how weathering might affect rocks over a very long period of time.

Erosion and Deposition

Weathering produces sediment. Sediment is loose material. It can be large boulders or tiny grains of sand. What happens to these weathered pieces of rock? Often they are moved to a new place. The moving of sediment from one place to another is the second step in the process, **erosion.** Wind, water, ice, living things, and gravity all can move sediment from one place to another.

DCI ESS2.A: Earth Materials and Systems. Rainfall helps to shape the land and affects the types of living things found in a region. Water, ice, wind, living organisms, and gravity break rocks, soils, and sediments into smaller particles and move them around. (4-ESS2-1)
CCC Cause and Effect. Cause and effect relationships are routinely identified, tested, and used to explain change. (4-ESS2-1)

What happens to the sediment that is carried away? It is deposited or dropped in a new place. This action is another process called **deposition.** Moving water can deposit pebbles and sand downstream. Or wind can deposit sand on a beach to form a sand dune. Together the three processes of weathering, erosion, and deposition can shape and change the land.

Long ago, on what is now Deer Isle, Maine, huge moving masses of ice carried these boulders from far away. When the ice melted, the boulders were left in a new place.

Wrap It Up! My Science Notebook

1. **Recall** What are some causes of erosion?
2. **Generalize** How do the effects of erosion and deposition change Earth's surface?
3. **Conclude** Why does erosion occur before deposition?

Wind Changes the Land

Wind can shape land through weathering, erosion, and deposition. Wind picks up and moves small pieces of sediment, such as soil and sand. When sand slams into rocks, tiny particles of the rocks break off. The moving sand acts like sandpaper, rubbing the rocks smooth.

Sand dunes form in windy places where sand is plentiful, such as on a beach or in a desert. Steady winds push grains

If conditions are right, grains of sand collect and form a sand dune. A sand dune is a landform caused when wind deposits sand.

of sand along the ground. Large objects such as rocks and plants block the sand's movement. A sand dune forms as more and more grains of sand are blocked by an object and become trapped. Over time, a sand dune may grow very tall. How tall? A few sand dunes reach heights of more than 500 meters (1640 feet)!

These landforms are called toadstool caprocks. The rock at the top has resisted weathering more than the rock beneath. The softer, lower rocks have worn away, leaving the unusual shapes.

Wrap It Up!

1. **Describe** How can wind change the shape of rocks?

2. **Cause and Effect** Describe the actions involved in the formation of sand dunes.

3. **Classify** Identify these two processes as erosion or deposition: Sand dunes form. Wind picks up particles of soil and sand.

Water Changes the Land

Moving water can change Earth's surface in a big way. For example, it took over a million years for the Virgin River to form the Zion Canyon in Utah. How does moving water change Earth's surface? Like wind, moving water carries sediment, such as sand and pebbles.

DCI ESS2.A: Earth Materials and Systems. Rainfall helps to shape the land and affects the types of living things found in a region. Water, ice, wind, living organisms, and gravity break rocks, soils, and sediments into smaller particles and move them around. (4-ESS2-1)
CCC Cause and Effect. Cause and effect relationships are routinely identified, tested, and used to explain change. (4-ESS2-1)

Sediment carried by the Virgin River scraped and chipped away rock on the sides and bottom of the river. It took many years for the river to carve the canyon.

If you hiked through the canyon, you might see that many of the rocks are smooth and rounded. Sediment carried along by the water bumps and rubs against larger rocks. Eventually the surface of the rocks wears away. The rocks become smoother, more rounded, and smaller. Weathering, erosion, and deposition are mostly slow processes, but over time they can make huge changes to Earth's surface.

Over time, the Virgin River eroded enough rock to form this canyon.

Rushing water in a stream in Zion Canyon weathers and erodes rocks, making them smoother.

Wrap It Up!

1. **Explain** How can water cause weathering, erosion, and deposition of rock?

2. **Generalize** How are some canyons a result of changes to Earth's surface by water?

Investigate

Weathering and Erosion

? How can you model the processes of weathering and erosion?

Scientists use various kinds of models to investigate how natural processes work. In this activity you'll explore two ways that weathering and erosion can change a rock called sandstone.

Materials

sandstone	paper towel	jar with lid
water	hand lens	stopwatch

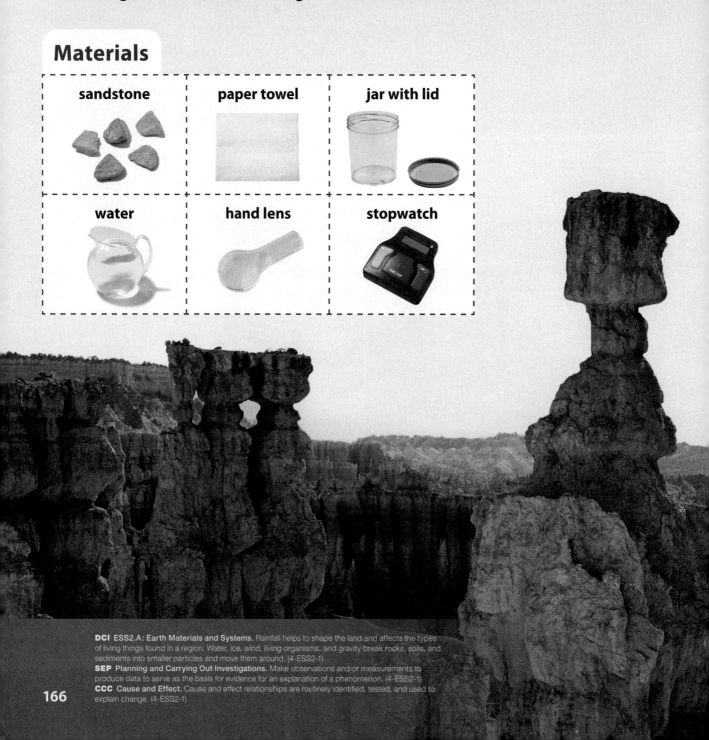

DCI ESS2.A: **Earth Materials and Systems.** Rainfall helps to shape the land and affects the types of living things found in a region. Water, ice, wind, living organisms, and gravity break rocks, soils, and sediments into smaller particles and move them around. (4-ESS2-1)
SEP Planning and Carrying Out Investigations. Make observations and/or measurements to produce data to serve as the basis for evidence for an explanation of a phenomenon. (4-ESS2-1)
CCC Cause and Effect. Cause and effect relationships are routinely identified, tested, and used to explain change. (4-ESS2-1)

1 Predict what will happen when you rub two pieces of sandstone together. Record your prediction in your science notebook. Hold the pieces of sandstone over the paper towel. Rub them together for a few seconds. Observe what happens. Record your observations.

2 Place five pieces of sandstone in a jar. Pour water into the jar to fill it about half full. Put the lid on the jar securely. Use the hand lens to observe the sandstone and the bottom of the jar. Record your observations.

3 Predict what will happen if you shake the jar. Then shake the jar for 3 minutes. Use the stopwatch to time yourself.

4 Observe the sandstone and the bottom of the jar. Record your observations.

Look at the photo. How might weathering and erosion have changed these rocks?

Wrap It Up!

1. **Compare** How do your results compare to your predictions? Explain.

2. **Analyze** What processes did you model in steps 1 and 3? Explain.

3. **Apply** Use what you learned in this investigation to explain the causes behind the effects on the rocks in the photograph.

Ice Changes the Land

What can move rocks that are as big as houses or as small as silt or sand? Ice! A **glacier** is a huge area of slow-moving ice. In some places, massive glaciers changed Earth's surface as they slowly moved over the land. As glaciers moved, they scraped against rock, carving it into new shapes.

Ice can change Earth's surface in another way, too. Water seeps into cracks in rock. The water freezes and expands. The ice inside the cracks acts like a wedge. It makes the cracks wider. The repeated freezing and thawing of ice will eventually widen the crack so much that the rock splits apart.

Long ago, a glacier scraped this rock smooth.

Glaciers carved wide, U-shaped valleys in Yosemite National Park in California.

This hanging valley formed when a small, shallow glacier flowed into a larger glacier.

Glaciers carved steep cliffs in Yosemite Valley.

Wrap It Up! My Science Notebook

1. **Compare** How do water and ice affect rock in similar ways?

2. **Explain** How does Yosemite Valley show that ice changes Earth's surface?

Living Things Change the Land

Organisms, or living things, can change the shape of the land. Over time, plants, animals, and other kinds of organisms can break rocks into smaller pieces and move them from one place to another.

Some organisms move rocks and sediments to new places. The roots of most plants grow in soil. As they grow, they push the particles of soil aside.

Meerkats are excellent diggers, often making huge underground burrows. This meerkat is digging for insects in the Kalahari Desert of South Africa.

DCI ESS2.A: Earth Materials and Systems. Rainfall helps to shape the land and affects the types of living things found in a region. Water, ice, wind, living organisms, and gravity break rocks, soils, and sediments into smaller particles and move them around. (4-ESS2-1)
DCI ESS2.E: Biogeology. Living things affect the physical characteristics of their regions. (4-ESS2-1)
CCC Cause and Effect. Cause and effect relationships are routinely identified, tested, and used to explain change. (4-ESS2-1)

Animals such as moles, gophers, worms, and many kinds of insects burrow through the ground. To make their burrows, they move particles of soil. A large animal such as a woodchuck can even move rocks!

Some animals may also use soil or rocks as building materials. A beaver uses soil, rocks, and sticks to make a hollow lodge. It also uses these materials to build a dam across a stream or river. Some birds use mud to build their nests. A robin may make hundreds of trips to collect bits of mud for its nest!

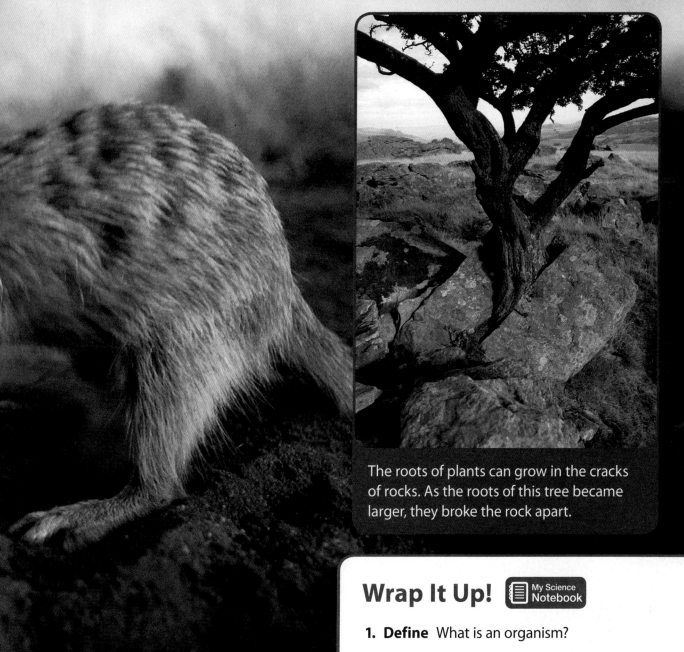

The roots of plants can grow in the cracks of rocks. As the roots of this tree became larger, they broke the rock apart.

Wrap It Up!

1. **Define** What is an organism?
2. **Cause and Effect** How can plants break down rocks?

Landslides Change Earth's Surface

Gravity is the force that pulls objects toward Earth's center. Gravity plays a part in the erosion of land. Have you ever seen a road sign that says *Caution: Falling Rock*? Rock on a ledge or steep hill may suddenly break loose and fall. Often it's just a rock or two. But at other times, it's a landslide.

This landslide in El Salvador destroyed many buildings in the neighborhood at the bottom of the hill.

A **landslide** is a rapid movement of rock, soil, and other material down a hill or mountain. A landslide can change Earth's surface. If enough material falls during a landslide, it may cause changes in the shape of the hill or mountain.

What causes landslides? Sometimes heavy rains can loosen soil and rock on slopes. Volcanoes and earthquakes can also start the motion. But it is the force of gravity that pulls the rocks and other loosened material downhill.

Landslides can leave large, bare areas on mountains and hills.

Wrap It Up! My Science Notebook

1. **Explain** How do landslides cause rapid changes to Earth's surface?

2. **Cause and Effect** What makes material in any landslide move downhill?

3. **Infer** How might a landslide affect people?

NATIONAL GEOGRAPHIC | Think Like an Engineer

Make Observations

You have read how running water can change the shape of the land. Now you'll use what you've learned to help solve a problem.

After a heavy rainstorm, a farmer discovered water making gullies in the soil. Water was eroding the soil as it made its way from a field to the nearby stream. If this continued, much of the soil would be washed away. Then the farmer would not be able to grow crops in the widening gully.

You have been asked to solve the problem of soil eroding from the farmer's field. To test your solution, you'll use small-scale models to investigate the rate, or speed, of erosion.

1. **Define the problem.** *My Science Notebook*
   How can you slow or prevent erosion of a farmer's field?

2. **Find a solution.**
   - To solve this problem, you will need to make a model of a hillside. You need to decide on the materials you will need for this model.

   - After you build your hillside, you need to decide how you will measure the rate of erosion. What are some ways you could change how water flows down your model hillside? Which variables will you change when you test your model? Which variables will you keep the same? How will you know the tests you run are fair?

   - Write out a plan for your investigation. Include a diagram of your hillside and the solutions to reduce erosion that you will test. Then collect the materials you need and build your model.

PE 4-ESS2-1. Make observations and/or measurements to provide evidence of the effects of weathering or the rate of erosion by water, ice, wind, or vegetation.
PE 3-5-ETS1-3. Plan and carry out fair tests in which variables are controlled and failure points are considered to identify aspects of a model or prototype that can be improved.

Too much rain can erode the soil in which crops are grown.

Think Like an Engineer
(continued)

3. **Test your solution.**
   - Carry out the test you planned. Collect data from your test.
   - Repeat your test several times, changing the variables you identified in your plan. Record your observations.

4. **Refine or change your solution.**
   - Analyze your data. Did your solution change the rate of erosion? Did each trial produce similar results? Did changing the variables you identified in your plan produce a better result? Could you make your solution work better? Test your new ideas.
   - When you are satisfied with your solution, get ready to explain your plan to others. Draw a detailed diagram or make a poster of your solution. Include clear labels that explain how your solution will work. Be prepared to defend your solution as others question or challenge your plan. Plan your argument in detail. Be sure to provide evidence from your tests to support your solution.
   - Share your solution with a partner. Record his or her comments and suggestions. Evaluate your partner's comments. Which of the suggestions would improve your plan? Revise your plan to include the suggestions.

5. **Present your solution.**
   When you are satisfied that your solution is the best you can make it, revise your presentation. Then share your solution with the class.

Farmers prepare their fields in certain ways to prevent soil erosion. They plough the crop rows in directions that block water from running straight downhill. That gives water more time to soak into the soil. They also alternate sections of crops. Rows of sturdy plants, less likely to be washed away by erosion, protect more delicate plants.

Natural Hazards

Natural processes are constantly changing Earth's surface. In some cases, these processes are harmful to humans and other living things. Something that is harmful or dangerous is called a **hazard.** Earthquakes, tsunamis, and volcanoes are three natural processes that can be hazardous to humans.

Earthquakes

An **earthquake** is the shaking of Earth's surface caused by sudden movement of rock beneath the surface. Earthquakes can cause buildings to collapse and roads and bridges to buckle. They may also break power lines and water pipes.

Volcanoes

Volcanoes form when molten rock from deep inside Earth rises to the surface. Volcanic eruptions can spew hot ash and molten rock high into the air. When these materials come down, they can bury buildings and roads and damage crops. Volcanic ash in the air can also disrupt airline traffic.

Tsunamis

A **tsunami** is a series of ocean waves caused by an underwater earthquake, an underwater volcanic eruption, or a landslide. Anything that causes Earth's surface to move beneath the water also moves the water. When large tsunamis come ashore, they can destroy buildings, roads, or even entire villages.

Wrap It Up!

1. **Define** What does the word *hazard* mean?

2. **List** What are three of the natural processes that can be hazardous to humans?

3. **Infer** Why might a volcanic eruption be dangerous for airline traffic?

Earthquakes

An earthquake is a natural process caused by the movement of parts of Earth's surface. Earthquakes start along a fault boundary. A **fault** is a break in Earth's surface where huge slabs of rock slip past, move away from, or push against each other. The slabs of rock often become locked together along a fault line. If the slabs break free, energy is released and moves through the rocks. This makes the ground shake.

Several million earthquakes occur every year. Most of them happen far from people or are too weak for people to notice.

Strong earthquakes can be very dangerous. Their violent shaking can raise and lower the land and change the course of rivers. Powerful earthquakes can damage buildings and other structures. Roads buckle, railroad tracks twist, and bridges collapse. Water pipes and electric power lines break. It can take years for people to repair the damage caused by a strong earthquake.

An earthquake near Oakland, California, caused this section of freeway to fall apart.

DCI ESS3.B: Natural Hazards. A variety of hazards result from natural processes (e.g., earthquakes, tsunamis, volcanic eruptions). Humans cannot eliminate the hazards but can take steps to reduce their impacts. (4-ESS3-2)
CCC Cause and Effect. Cause and effect relationships are routinely identified, tested, and used to explain change. (4-ESS3-2)

Wrap It Up!

1. **Define** What is a fault?

2. **Cause and Effect** How can an earthquake affect structures built by humans? Give three examples.

3. **Infer** How might an earthquake affect people's ability to get from place to place?

Investigate

Earthquakes

? **How can you demonstrate liquefaction?**

During earthquakes, many buildings have collapsed or tilted and sunk into the ground. Why? Often the buildings were built on loose materials such as rocky soil, sand, or mud. There is also a lot of water in the ground. When the earthquake shook the ground, these materials became liquid-like. The change of a solid area of ground such as sand or mud into a less stable liquid-like condition is called **liquefaction.** Liquefaction is one of the hazards caused by earthquakes.

In this investigation, you will use a model to learn how earthquakes affect structures built on sand or mud.

Materials

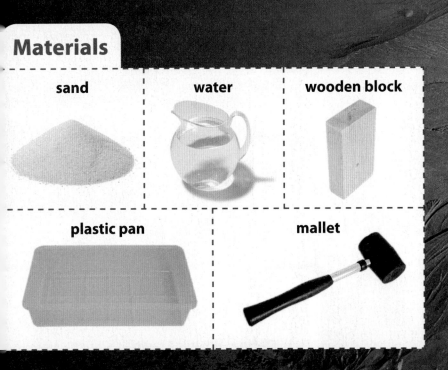

sand

water

wooden block

plastic pan

mallet

DCI ESS3.B: Natural Hazards. A variety of hazards result from natural processes (e.g., earthquakes, tsunamis, volcanic eruptions). Humans cannot eliminate the hazards but can take steps to reduce their impacts. (4-ESS3-2)
CCC Cause and Effect. Cause and effect relationships are routinely identified, tested, and used to explain change. (4-ESS3-2)

1 Fill the pan with sand, leaving about 9 centimeters at the top.

2 Place the pan on a table. Then pour in water to just below the surface of the sand. Record your observations.

3 Push one end of the block down into the wet sand so it stands up like a building. Predict what will happen when you repeatedly tap the mallet against the pan. Record your prediction.

4 Hold the pan in place. Very gently tap the side of the pan repeatedly with the mallet. Observe what happens to the sand and the block. Record your observations.

The water-filled holes in this photograph were caused as water bubbled up through the soil as a result of an earthquake. Imagine how such a change in the soil would affect a building!

Wrap It Up!

1. **Explain** What did the block represent? What did hitting the pan with the mallet represent?

2. **Cause and Effect** How did this investigation demonstrate the effects of liquefaction?

3. **Revise** How could you change the model to test ways to make buildings constructed on sand or mud more stable? Write your plan. Include a diagram of how your test would work.

Tsunamis

A tsunami is a series of fast-moving ocean waves caused by an earthquake, or an underwater volcanic eruption, or a landslide. In places where the ocean is deep, tsunami waves may be only a few centimeters high. As the waves approach shore, they increase in height. Some tsunami waves come ashore gently. Others become huge walls of water.

The Great Tōhoku Earthquake of 2011 in Japan set off a series of tsunamis. A few minutes after the earthquake, one tsunami crashed over this seawall in Miyako City. Tsunamis up to 12 meters (39 feet) tall destroyed coastal areas across northeastern Japan.

When a large tsunami smashes into land, its turbulent water can destroy houses, roads, and other structures. People may be swept away. Very large tsunamis can destroy entire villages.

This tsunami was caused by a strong earthquake that struck Japan in March of 2011.

Wrap It Up!

1. **Recall** What are three events that cause tsunamis?

2. **Explain** How do some tsunamis change when they reach land?

3. **Summarize** Why are tsunamis dangerous?

Volcanoes

A volcanic eruption is a natural process on Earth's surface. Volcanoes form when **magma,** or melted rock inside Earth, rises to the surface. In all volcanoes, molten rock **erupts** or flows through an opening. Magma that erupts onto Earth's surface is called **lava.** Sometimes the lava flows down the side of the volcano and hardens into rock.

Mt. Sakurajima on the main island of Japan erupted on August 18, 2013.

DCI ESS3.B: Natural Hazards. A variety of hazards result from natural processes (e.g., earthquakes, tsunamis, volcanic eruptions). Humans cannot eliminate the hazards but can take steps to reduce their impacts. (4-ESS3-2)

Hot, expanding gases in magma can cause explosive eruptions. Hot ash and lava released during an eruption may quickly bury or destroy nearby forests, fields, and towns. Such eruptions are a hazard to the people living nearby. The ash in the air makes it difficult to breathe. Volcanoes may also release poisonous gases.

There is little people can do to limit the dangers of a volcano. The safest solution is to not live close to one.

Ash covered these rental cars in Kagoshima, Japan, after Mt. Sakurajima erupted.

Wrap It Up!

1. **Identify** What are some of the hazards caused by erupting volcanoes?

2. **Compare and Contrast** What is the difference between magma and lava?

3. **Research** Where is the nearest volcano to your region? When did it last erupt?

Reducing the Impact of Natural Hazards

A natural hazard may have large impacts on the environment or on people's lives. You have read about some of the impacts caused by earthquakes, tsunamis, and volcanoes. When a natural hazard causes great damage or loss of life, it becomes a natural disaster.

Natural hazards cannot be eliminated, but people can take steps to reduce their impacts. For example, engineers can design buildings and bridges to withstand the violent shaking of earthquakes. This reduces damage and saves lives.

DCI ESS3.B: **Natural Hazards.** A variety of hazards result from natural processes (e.g., earthquakes, tsunamis, volcanic eruptions). Humans cannot eliminate the hazards but can take steps to reduce their impacts. (4-ESS3-2)
CETS Influence of Engineering, Technology, and Science on Society and the Natural World. Engineers improve existing technologies or develop new ones to increase their benefits, to decrease known risks, and to meet societal demands. (4-ESS3-2)

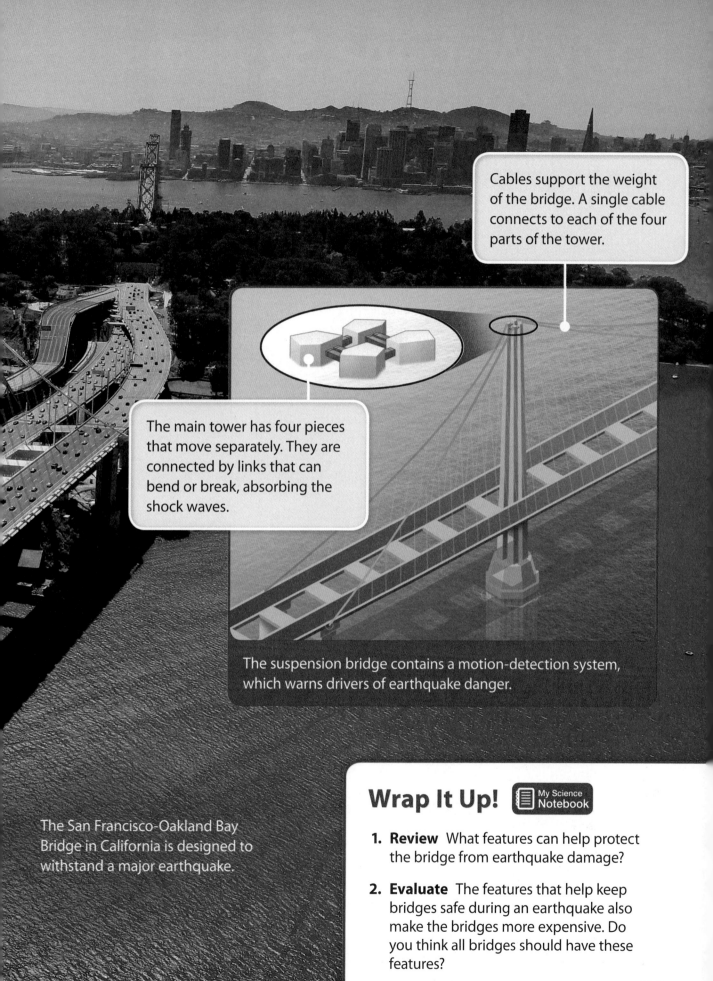

Cables support the weight of the bridge. A single cable connects to each of the four parts of the tower.

The main tower has four pieces that move separately. They are connected by links that can bend or break, absorbing the shock waves.

The suspension bridge contains a motion-detection system, which warns drivers of earthquake danger.

The San Francisco-Oakland Bay Bridge in California is designed to withstand a major earthquake.

Wrap It Up!

1. **Review** What features can help protect the bridge from earthquake damage?

2. **Evaluate** The features that help keep bridges safe during an earthquake also make the bridges more expensive. Do you think all bridges should have these features?

Early Warning Systems

If people know that a natural hazard is likely to occur, they can take steps to reduce its impact. But how can people know when they are in danger? Scientists study earthquakes, tsunamis, and volcanoes to find patterns for their causes. They use what they learn to better predict when natural hazards are likely to happen. Engineers apply what scientists learn to design tools that help predict natural hazards and warn people about them.

Seismometers measure earthquake activity. An increase in small earthquakes can mean a larger earthquake is about to happen or a volcano is becoming more active. If a volcanic eruption is predicted, people can **evacuate,** or move to safer areas. Governments can prepare emergency supplies, such as food and water. Emergency workers can be ready to help with injuries. All of these steps reduce the impact and help save lives.

Seismometers detect earthquake waves. An increase in earthquakes near a volcano can be a sign of increasing volcanic activity.

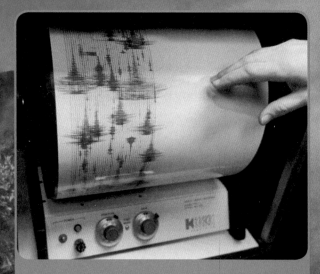

Ground motions sensed by seismometers are converted into electronic signals. The signals are transmitted by radio and recorded on **seismographs.**

DCI ESS3.B: Natural Hazards. A variety of hazards result from natural processes (e.g., earthquakes, tsunamis, volcanic eruptions). Humans cannot eliminate the hazards but can take steps to reduce their impacts. (4-ESS3-2)
CCC Cause and Effect. Cause and effect relationships are routinely identified, tested, and used to explain change. (4-ESS3-2)
CETS Influence of Engineering, Technology, and Science on Society and the Natural World. Engineers improve existing technologies or develop new ones to increase their benefits, to decrease known risks, and to meet societal demands. (4-ESS3-2)

Scientists use several devices and methods to constantly collect information about a volcano's activity. Radio transmissions allow the data to be sent instantly to a monitoring system. Scientists can tell what is happening with a volcano almost as soon as it happens. They can also use that information to predict what is likely to happen next.

In addition to tracking seismic activity, scientists also monitor gases, temperature, and water at a volcano site. A change in the amount of carbon dioxide and other gases given off by the volcano can signal a coming eruption. So can temperature changes in rocks at the surface or rocks underground. Scientists monitor water levels and look for changes in chemistry in water near volcanoes, too.

Swelling, sinking, or cracking of the ground near a volcano can mean magma is moving beneath the surface. Tiltmeters detect any movement of magma close to the surface.

Scientists continually monitor the gases coming out of active volcanoes.

Wrap It Up!

1. **Describe** What information do scientists collect to monitor volcanoes?

2. **Explain** How is the information used to reduce the impact of volcanic eruptions?

Tsunami Detection

Most of the events that cause tsunamis occur on the floor of the ocean. It's not easy to collect seismic activity data there. It's also not possible to monitor the entire ocean floor. A sudden earthquake or landslide on the sea floor can cause a tsunami that may strike land hundreds or even thousands of kilometers away.

Engineers have designed a system of floating devices called buoys that scientists use to measure changes in the depth of ocean water. A pattern of depth changes can indicate that a tsunami has formed. If scientists can detect a tsunami as soon as it happens, they can alert people on land that a tsunami may be on the way. People can then leave areas near the coast and move inland or to higher ground.

The tsunami warning buoy is floating on the ocean.

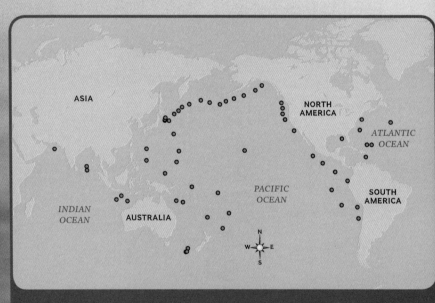

The map shows the ring of tsunami warning buoys that encircles the Pacific Ocean.

DCI ESS3.B: **Natural Hazards.** A variety of hazards result from natural processes (e.g., earthquakes, tsunamis, volcanic eruptions). Humans cannot eliminate the hazards but can take steps to reduce their impacts. (4-ESS3-2)
CETS Influence of Engineering, Technology, and Science on Society and the Natural World. Engineers improve existing technologies or develop new ones to increase their benefits, to decrease known risks, and to meet societal demands. (4-ESS3-2)

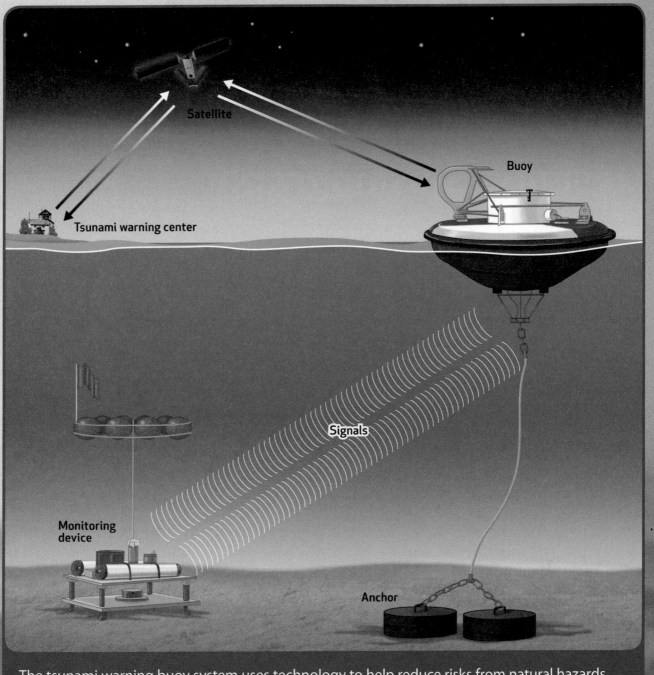

The tsunami warning buoy system uses technology to help reduce risks from natural hazards. Buoys receive data from monitoring devices on the sea floor. When the system of buoys detects a tsunami, a satellite transmits a signal to a tsunami warning center. The warning center then alerts people that a tsunami is approaching. In some cases, loud horns sound the alarm. Then people can quickly move away from the coast.

Wrap It Up!

1. **Describe** How do scientists detect tsunamis?
2. **Explain** Why are satellites useful in predicting tsunamis?

Design a Seismograph

During an earthquake, the ground can shake violently. This shaking can damage buildings and roads. People can also get hurt. You have read about some tools scientists use to predict and measure earthquakes. A seismograph is a tool that records up-and-down and side-to-side patterns of ground motion. The patterns show how strong an earthquake is and how long it lasts.

One seismograph design uses a heavy weight that hangs from a rigid frame. A pen at the bottom of the weight touches a moving roll of paper. When the earth shakes, the pen records a pattern of shaking on the paper.

You and your engineering team will make a model seismograph. You will use it to measure the strength of shaking caused by an "earthquake" you set off.

How will the earthquake damage you see here affect the people living in this area?

The Challenge

Your challenge is to design and build a model of a seismograph. Your seismograph must:

- record data without stopping for 5 seconds
- record shaking from two different locations
- show two different patterns of shaking

STEM
ENGINEERING PROJECT
(continued)

A basic tool for detecting earthquakes was invented in China nearly 2,000 years ago.

1 Define the problem.

Think about the problem you are solving. What does your model seismograph need to do? The goals it must meet are the criteria of the problem. You will know your design is successful if your model meets the criteria. Look back at the challenge box. The list describes the criteria for your model.

Constraints are things that limit your design. For example, you can only build your model using materials your teacher gives you. Other limits might include the time you have to build or the space available for your model.

Write the problem you need to solve in your notebook. List the criteria and constraints.

2 Find a solution.

Your teacher will give you directions for making the part of your seismograph that records patterns of shaking. It is shown below. Make your recording device, and practice using it.

Next your teacher will give you materials to build a frame for your seismograph. It must support your recording device. It must have a way to smoothly feed paper to record patterns. Each person on your team should sketch a design and share ideas. Together choose a design you think is best. Make a sketch of the final design.

recording device

Scientists think this tool was designed so that an earthquake would cause the rod to swing. The movement would trigger a ball to drop from one of the dragon's mouths. Observing which toad "caught" the ball revealed the direction in which the earthquake was happening.

When this rod swung, its motion would cause a ball to drop into one of the cups below.

3 Test your solution.

Have your teacher approve the design for your seismograph. Then build your model. Is it strong enough to hold the recording device? Does the recording device touch the paper? Adjust your seismograph to make it work. Make notes to show what you change.

Test your model. Set it on a table. Make an "earthquake" by shaking the table side to side as a team member runs the paper through the seismograph. Remember to time your test. Repeat your test. Next, choose a way to create up-and-down shaking movements. Record your results.

Check the patterns you recorded. Did your seismograph record both kinds of ground motion caused by the "earthquake"? Does it show a different pattern for each?

Discuss the results of the tests with your team. Did your model meet the criteria of the problem?

4 Refine or change your solution.

Talk with your team about how you can improve your seismograph. Use your ideas to make changes. Then test your seismograph the same way you did in Step 3. Did your changes make a difference? How do you know?

Present your model seismograph to the class. Explain the results of your tests. Show your recordings. Explain which patterns show a strong shaking and which show weaker shaking. Answer questions about your design. Ask questions about the other teams' seismographs.

Think about the results of your tests and the feedback from the class. How could you improve your design? Record your ideas in your notebook.

Patterns of Water and Land Features

Maps can be used to find the location of land and water features on Earth. This map shows some of the major features on land, such as mountain ranges or chains.

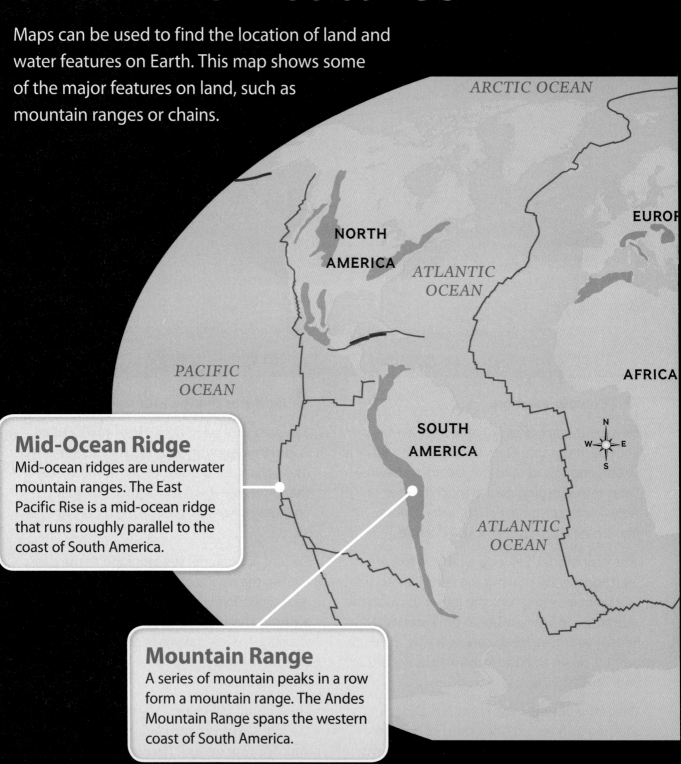

Mid-Ocean Ridge
Mid-ocean ridges are underwater mountain ranges. The East Pacific Rise is a mid-ocean ridge that runs roughly parallel to the coast of South America.

Mountain Range
A series of mountain peaks in a row form a mountain range. The Andes Mountain Range spans the western coast of South America.

DCI ESS2.B: Plate Tectonics and Large-Scale System Interactions. The locations of mountain ranges, deep ocean trenches, ocean floor structures, earthquakes, and volcanoes occur in patterns. Most earthquakes and volcanoes occur in bands that are often along the boundaries between continents and oceans. Major mountain chains form inside continents or near their edges. Maps can help locate the different land and water features areas of Earth. (4-ESS2-2)
SEP Analyzing and Interpreting Data. Analyze and interpret data to make sense of phenomena using logical reasoning. (4-ESS2-2)

made up of thousands of volcanic peaks. A **deep ocean trench** is a steep underwater canyon.

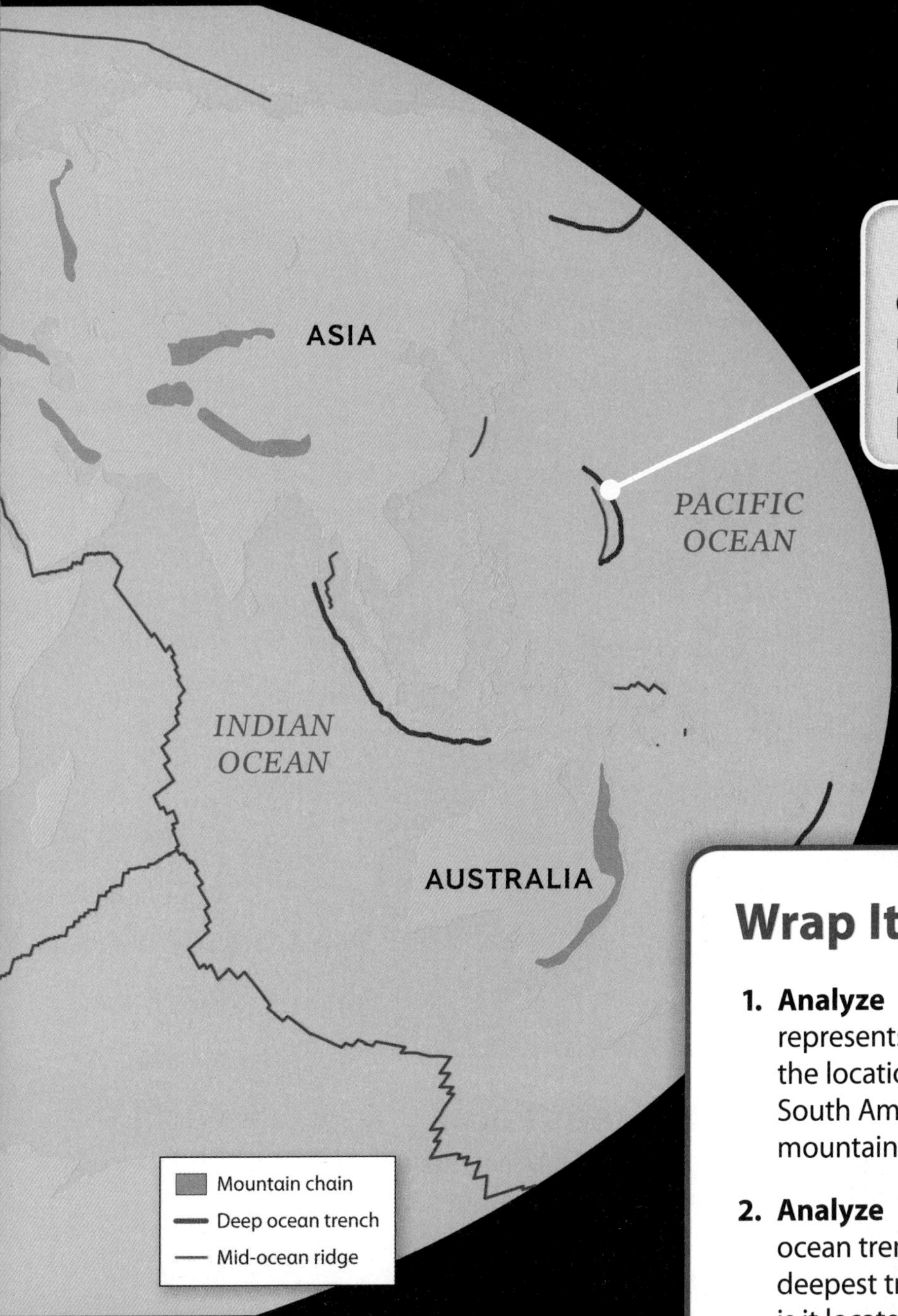

Deep Ocean Trench
Ocean trenches are steep underwater canyons. The Mariana Trench is the deepest place in the world's ocean.

ASIA

PACIFIC OCEAN

INDIAN OCEAN

AUSTRALIA

Mountain chain
Deep ocean trench
Mid-ocean ridge

Wrap It Up! My Science Notebook

1. **Analyze** What symbol on the map represents a mountain range? How does the location of the mountain range in South America differ from that of the main mountain range in Asia?

2. **Analyze** What symbol represents a deep ocean trench? The Mariana Trench is the deepest trench on Earth. In which ocean is it located? Which continent is located to the west of this trench? *Hint*: Use the compass rose to find directions.

3. **Interpret** Describe the location of the East Pacific Rise.

NATIONAL GEOGRAPHIC | **Think Like a Scientist**

Analyze and Interpret Data

This is the same map as on the previous pages, but two features have been added to it. One of these is the location of Earth's active volcanoes. The other is the location of some of the major earthquakes that have occurred since 1900. Study the map to see what patterns you can find.

Hawai'i
The Hawaiian Islands consist of many volcanoes. Kīlauea, Mauna Loa, and Hualalai have erupted within the past 200 years.

San Andreas Fault
The San Andreas Fault extends through California for about 1,300 kilometers (about 800 miles). It is the site of frequent earthquakes.

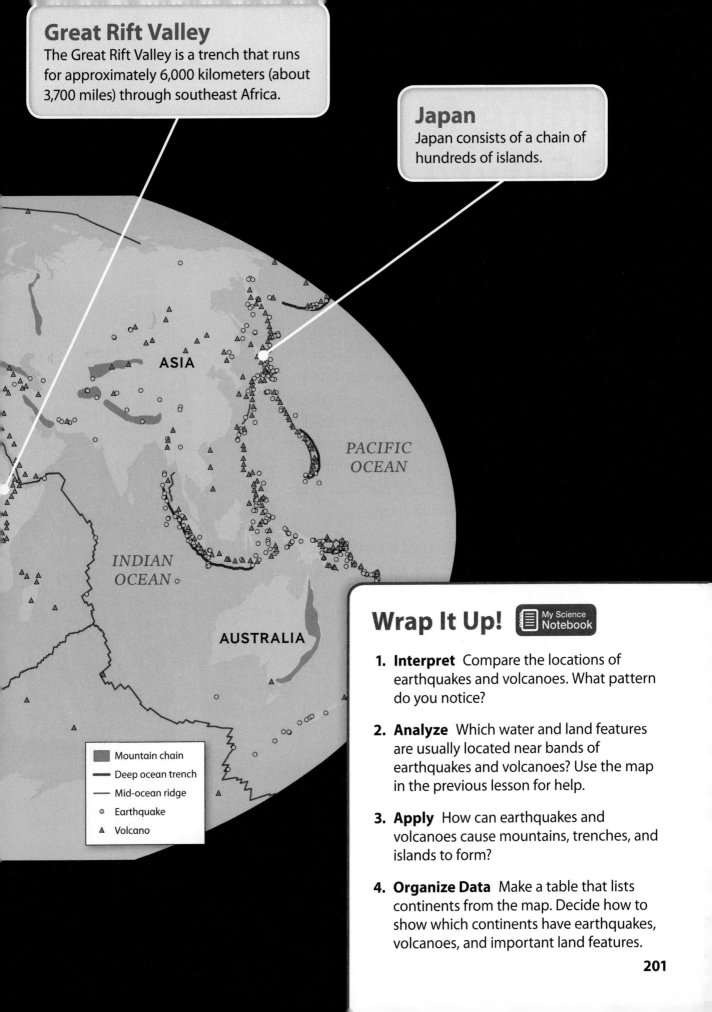

Great Rift Valley
The Great Rift Valley is a trench that runs for approximately 6,000 kilometers (about 3,700 miles) through southeast Africa.

Japan
Japan consists of a chain of hundreds of islands.

Mountain chain
Deep ocean trench
Mid-ocean ridge
○ Earthquake
▲ Volcano

Wrap It Up!

1. **Interpret** Compare the locations of earthquakes and volcanoes. What pattern do you notice?

2. **Analyze** Which water and land features are usually located near bands of earthquakes and volcanoes? Use the map in the previous lesson for help.

3. **Apply** How can earthquakes and volcanoes cause mountains, trenches, and islands to form?

4. **Organize Data** Make a table that lists continents from the map. Decide how to show which continents have earthquakes, volcanoes, and important land features.

NATIONAL GEOGRAPHIC | Think Like an Engineer
Case Study

Building for the Future

Problem

How can engineers make buildings more earthquake resistant?

In March 2011 a powerful earthquake just off the coast of Japan shook the ground and set off an enormous tsunami. The earthquake and tsunami damaged more than one million buildings and killed or injured thousands of people. Yet most buildings survived the earthquake, and many lives were saved. Why was the damage in Japan less than might have been expected? Because earthquakes are common in Japan, the government requires all new buildings to be designed and built to withstand earthquakes.

The disaster inspired an engineer named Masaaki Saruta and his team to develop structures that are even more resilient. The engineers knew they could not prevent earthquakes, but they could develop solutions that would reduce the impact of these disasters.

This building was destroyed by an earthquake in Japan.

DCI ESS3.B: Natural Hazards. A variety of hazards result from natural processes (e.g., earthquakes, tsunamis, volcanic eruptions). Humans cannot eliminate the hazards but can take steps to reduce their impacts. (4-ESS3-2)
DCI ETS1.B: Designing Solutions to Engineering Problems. Testing a solution involves investigating how well it performs under a range of likely conditions. (secondary to 4-ESS3-2)
CETS Influence of Engineering, Technology, and Science on Society and the Natural World. Engineers improve existing technologies or develop new ones to increase their benefits, to decrease known risks, and to meet societal demands. (4-ESS3-2)

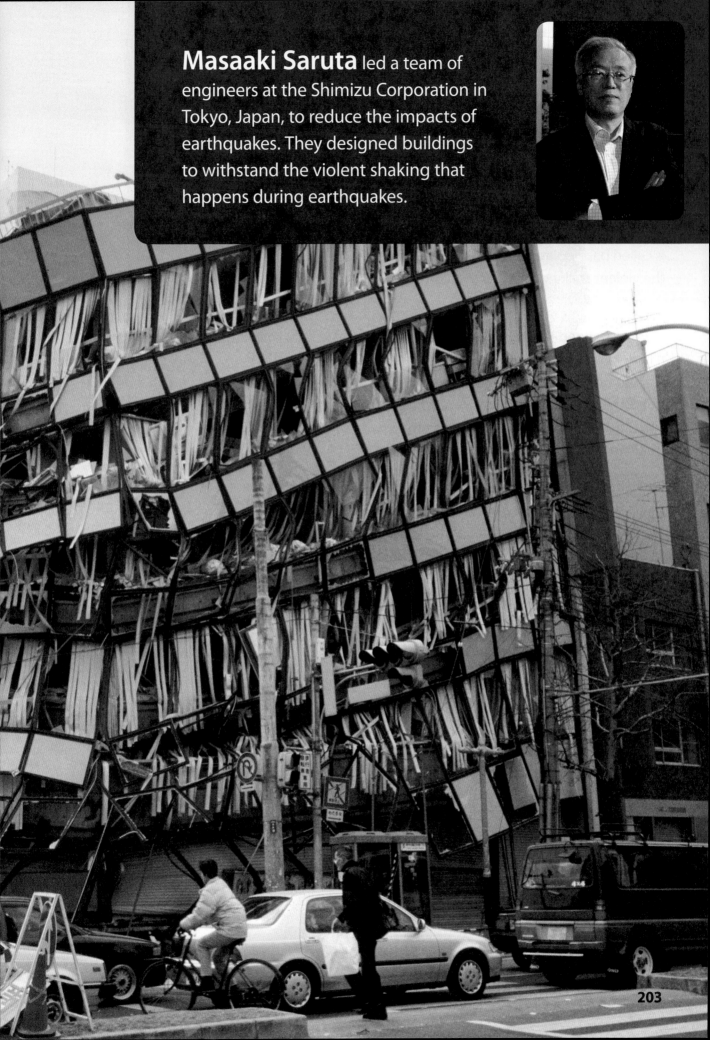

Masaaki Saruta led a team of engineers at the Shimizu Corporation in Tokyo, Japan, to reduce the impacts of earthquakes. They designed buildings to withstand the violent shaking that happens during earthquakes.

Think Like an Engineer
Case Study (continued)

Solution

Over the years, the engineers at Shimizu have developed a variety of earthquake-resistant designs. In some of the designs, buildings are separated from their foundations by a system of pads or bearings. This is called base isolation, and it separates the frame of a building from the violent shaking of an earthquake.

Masaaki and his team invented a new kind of base isolation called the core-suspended isolation system (CSI). In CSI, the core of a building is a large pillar of reinforced concrete. The floors of the building hang from the top of the core. The core is isolated from the rest of the building by large rubber bearings. When an earthquake strikes, the core absorbs the vibrations. The core may sway, but the floors of the building do not sway. Instead they remain upright.

The team has also designed a building that partially floats on water! The building stands on rubber bearings and its foundation rests in a pool of water. To test their design, they built a large model of the building. Their tests showed that water reduces the movement of a building. But the real test came during the March 2011 earthquake, when the engineers found that the features of their building cut the effects of shaking by more than half.

Masaaki said, "We want to come up with technologies that save people's lives." His team's engineering solutions have helped to accomplish this goal.

The Safety and Security Center was constructed using the core-suspended isolation system.

The building is suspended, or hangs, over a strong concrete core.

If the core shakes during an earthquake, layers of rubber bearings prevent much of the motion from being transferred to the building. This reduces damage.

Wrap It Up!

1. **Summarize** Describe how a building is constructed using the core-suspended isolation system.

2. **Explain** How does CSI reduce the impact of an earthquake?

3. **Relate** How are base isolation and core-suspended isolation systems related?

NATIONAL GEOGRAPHIC | Think Like an Engineer

Generate and Compare Solutions

You've read about Masaaki Saruta and his team's work to design and build earthquake-resistant buildings. Now it's your turn to use some of the same techniques engineers use. Working with a team, you'll design and test model houses to see if you can design and build a more damage-resistant structure.

1. **Define the problem.** 📓 My Science Notebook

    How can you make a house more earthquake resistant?

2. **Find a solution.**
    - Work with a team. Think about how earthquakes damage buildings. You may need to do some research to find out more about earthquake damage to different kinds of structures. Think about what kind of house you would build, how you would build it, and what kind of damage you would want to prevent. Record your ideas as you develop your plan.

    - Now think about methods you would use to make a house earthquake resistant. How could you determine whether the house was resistant to earthquakes? How could you simulate an earthquake? Write out a plan for an investigation. List the criteria for a successful design. Also list the constraints, or limits to your design.

3. **Conduct an investigation.**
    - Use materials your teacher provides to build a house. Subject the house to a simulated earthquake. What damage did it cause? How might you add to or change the building materials and the way the house is built to protect the house from damage?

PE 4-ESS3-2. Generate and compare multiple solutions to reduce the impacts of natural Earth processes on humans.
CETS Influence of Engineering, Technology, and Science on Society and the Natural World. Engineers improve existing technologies or develop new ones to increase their benefits, to decrease known risks, and to meet societal demands. (4-ESS3-2)

This model house has been built with a set of braces inside its structure to make it more stable in earthquakes. The design team uses a large moving platform to reproduce an earthquake. Then they check to see how well the model house has survived being shaken.

Think Like an Engineer
(continued)

- Build a new house. For this house, your teacher may provide additional building materials. Test your new house with a simulated earthquake. Record your observations.

- Adjust your design, and test again. You may choose to use a different building method or different materials. Be sure that you conduct a fair test. In a fair test you change only one variable each time you carry out the test. Record your observations each time you carry out a new test.

4. **Refine or revise your solution.**
   - Analyze your results. Which solutions worked best based on criteria and constraints? How do you know?

   - Once you have at least two solutions for making your house resistant to earthquake damage, develop a document or presentation that clearly describes your findings. Use diagrams with labels to explain your solutions, and provide evidence from your tests to support your selection of the best designs.

   - Share your designs with another team. Explain your evidence showing why you selected these two designs as the best ones. Record comments and suggestions from your classmates. Which of their ideas would improve your plan? After you have discussed your design solutions with your classmates, revise the design of your house to make it even more earthquake resistant. Compare and contrast your two designs by describing how they are alike and different from each other.

5. **Present your solution.**
   When you are satisfied that your solutions are the best you can make them, prepare a presentation. Then share your solutions with the class. Compare and contrast your solutions with those of other classmates.

The team examines cracks in the model to evaluate how well the construction resisted damage during the shaking.

The Badlands

These rock formations are found in Badlands National Park in South Dakota. The colorful horizontal bands are layers of different kinds of sedimentary rock. **Sedimentary rock** is formed from sediment on Earth's surface. The sediment is deposited in layers on land or on the sea floor. Over millions of years, the layers harden into rock. The oldest layers are on the bottom, and the newer layers are on the top.

Forces of wind and water erosion shape land across the Badlands region.

Each band of rock started out as a flat layer of sediments. The sediments included the remains of plants and animals. Some of those remains became **fossils,** or traces of plants and animals that lived long ago. Scientists use fossils to tell what kinds of organisms lived at the time the different layers of sediment were laid down. The location of certain fossils also indicates the order in which rock layers formed. Studying patterns of fossils and rock layers can help scientists understand how land changes over time due to earth forces.

DCI ESS1.C: The History of Planet Earth. Local, regional, and global patterns of rock formations reveal changes over time due to earth forces, such as earthquakes. The presence and location of certain fossil types indicate the order in which rock layers were formed. (4-ESS1-1)
CCC Patterns. Patterns can be used as evidence to support an explanation. (4-ESS1-1)
NS Scientific Knowledge Assumes an Order and Consistency in Natural Systems. Science assumes consistent patterns in natural systems. (4-ESS1-1)

30 MILLION YEARS AGO: The light-colored layers include sandstone and layers of ash from volcanoes. Fossils of desert plants show that the land was very dry.

33 MILLION YEARS AGO: The most colorful rocks were deposited when the climate was changing the land from a wet floodplain to drier grassland. Fossils found in these rock layers include tortoises.

37 MILLION YEARS AGO: The pale gray and green layers were deposited when there were forests, rivers, and shallow lakes on land. Fossils include enormous mammals that lived on savannahs.

65 MILLION YEARS AGO: The orange and yellow layers formed from mudflats and forest soils. These layers include clay-filled holes, which show that big trees grew in the soil.

75 MILLION YEARS AGO: Ammonite fossils are found in layers of limestone or shale and show that this area was once at the bottom of the ocean.

30 MILLION
33 MILLION
37 MILLION
65 MILLION
75 MILLION

Wrap It Up!

1. **Define** What is a fossil?

2. **Explain** How do the layers of sedimentary rock form? Where are the oldest layers usually found?

3. **Draw Conclusions** Scientists find two layers of sedimentary rock. One layer contains a fossil of a palm tree. The other layer contains an ammonite fossil. Which layer is older? Explain.

Iceland

Iceland is an amazing place. Although it is located in the cold North Atlantic Ocean, it has many hot springs, geysers, and 35 active volcanoes! Why does Iceland have so much volcanic activity? Because the island lies on top of the Mid-Atlantic Ridge, an underwater mountain chain made up of thousands of volcanic peaks. Iceland is the only place on Earth where the Mid-Atlantic Ridge rises above the surface of the ocean.

Over millions of years, volcanoes released thick flows of lava. The lava hardened into the dark rocks found all over Iceland. Volcanic rock is called igneous rock, and it does not form in layers. You won't find fossils in these rock formations, but their patterns do reveal how powerful forces have built up the land.

A large break in Earth's crust, called a **rift** or fault, runs through the middle of Iceland. It is part of the Mid-Atlantic Ridge. Land on each side of the rift pulls apart. Over time, this activity creates a new landform called a rift valley. Other land features can form, too. The lake in the photograph formed when volcanic activity caused the land to sink. Water then collected in the depression to form a lake.

Lake Thingvallavatn is located in Iceland's rift valley. The dark ridge and split in the land to the left of the lake show the fault where huge slabs of land are being pulled in opposite directions.

Wrap It Up!

1. **Identify** What pattern of natural activity explains why dark rocks are found all over Iceland?

2. **Cause and Effect** Why does Iceland have so much volcanic activity?

3. **Infer** What does Iceland's rift valley reveal about earth forces?

4. **Predict** What do you predict will happen to Iceland's rift valley in the future based on past activity of earth forces there?

NATIONAL GEOGRAPHIC | Think Like a Scientist

Identify Evidence

The Grand Canyon is an awesome wonder. Standing on its South Rim, you can see its plateaus and canyons spreading northward for about 29 kilometers (18 miles). In some places you can see the Colorado River at the bottom of the canyon, about 1,500 meters (5,000 feet) below.

The walls of the canyon are made up of horizontal layers of sedimentary rock. If you know what to look for, these layers reveal the history of the region. This is not the history of hundreds or thousands of years, but the history of how the landscape has changed over hundreds of millions of years!

The Colorado River carved the deep canyon over millions of years, exposing layer after layer of rock.

The different colored layers in the walls of the Grand Canyon contain different types of rock. The oldest layers are at the bottom. The newest layers are at the top.

NATIONAL GEOGRAPHIC | Think Like a Scientist
(continued)

Limestone, sandstone, and shale are all sedimentary rocks. Look at the patterns in the rock formations for evidence of how the landscape in this region has changed over time. Then cite evidence from the diagram to answer the questions.

SANDSTONE This layer formed when a desert with sand dunes covered the land. The tracks of spiders and scorpions were preserved in the sandstone.

REDWALL LIMESTONE These rocks formed in a wide, shallow sea. Sea animals such as corals and crinoids lived in the calm waters.

SHALE This layer formed when the land was covered by a warm, muddy sea. Animals such as trilobites, brachiopods, and crinoids lived here.

fossil trilobite

SCHIST This layer was once made of sedimentary rock. Forces inside Earth squeezed the rock until it changed into a fine-grained metamorphic rock called schist.

SANDSTONE This layer formed in an ancient sea where trilobites and brachiopods lived. There are traces of waves in the sands!

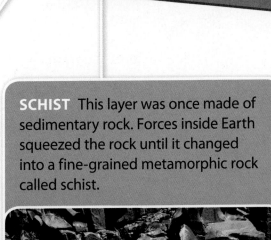

216

LIMESTONE This layer formed when the region was covered with a warm, shallow sea. Sponges, snails, corals, sharks, and fish lived in the clear water. This fossil shell was found in the limestone.

fossil gastropod

Wrap It Up!

1. **Interpret Diagrams** Which rocks make up the newest layer shown in the diagram? Which rocks make up the oldest layer?

2. **Sequence** Use evidence from the diagram and what you know about sedimentary rocks to order these events from first to last:

    a. Sand dunes cover the land.
    b. The Colorado River cuts through the land's layers, forming a deep canyon.
    c. Sedimentary rocks are squeezed together, forming schist.
    d. A warm, muddy sea covers the land.

3. **Infer** Would you expect to find fossils of coral in the sandstone layer here? Explain.

4. **Use Evidence** Explain how the landscape has changed over time. Use evidence such as the order of rock layers, fossils found in each layer, and the action of earth forces.

Stories in Science

A Feel for Fossils

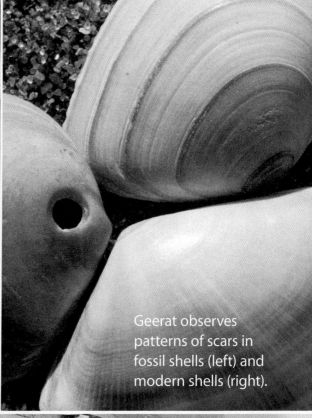

Geerat observes patterns of scars in fossil shells (left) and modern shells (right).

Geerat is shown studying shells in his lab. He also travels the world to work. As he says, "One cannot hope to understand nature without experiencing it firsthand."

CCC Patterns. Patterns can be used as evidence to support an explanation. (4-ESS1-1)
CCC Cause and Effect. Cause and effect relationships are routinely identified and used to explain change. (4-ESS3-1)
NS Scientific Knowledge Is Based on Empirical Evidence. Science findings are based on recognizing patterns. (4-PS4-1)
NS Scientific Knowledge Assumes an Order and Consistency in Natural Systems. Science assumes consistent patterns in natural systems. (4-ESS1-1)

"I was first shown some shells from Florida by my fourth-grade teacher. I was overwhelmed with their beauty." —Geerat Vermeij

Geerat Vermeij (GAIR-ē ver-MĀ) is a marine biologist. He studies shells. Yet he has never actually seen one. Doctors had to remove Geerat's eyes when he was three years old. Despite being blind, he became a scientist.

You may think a blind scientist is odd. After all, science is based on observations. "Observation is the first, and in many ways the most important, step in a scientific inquiry," Geerat says. He also points out that observations use all of a person's senses. Sight is just one of them.

As Geerat explains, "I listen and smell and feel." For example, he can tell the differences between shells by just touching them. He runs his fingers over shells, feeling for patterns. He feels their bumps, ridges, and scars.

Geerat studies how the designs of animals' shells have changed over time. He observes fossils collected from different rock layers. The fossils give him clues about what shelled animals looked like at different times in Earth's history. Geerat compares them to shells from today's animals.

Studying a shell's scars tells Geerat a story. They are evidence that the animal inside the shell was likely attacked by a predator. The predator tried to break the shell but could not. It could only make a crack or chip in the shell. It could not eat the animal inside. The animal's shell protected it.

Geerat has noticed a pattern in the shells he studies. From long ago until now, more and more shells have scars. More shelled animals are surviving because their shells protected them. Geerat says many of today's shells are thicker and stronger than shells of the past. This evidence shows that animals are building better shells than they did millions of years ago!

Wrap It Up!

1. **Explain** Why does Geerat Vermeij study fossils?

2. **Summarize** What has Geerat learned by studying shells of fossils and living animals? How did he reach his conclusions?

3. **Evaluate** Read Geerat's quote at the top of the page. How can a blind person be "overwhelmed by beauty"?

NATIONAL GEOGRAPHIC | Science Career

Crisis Mapper

What can help speed medical aid to areas destroyed by tsunamis? What can help direct rescue helicopters to areas struck by earthquakes? What can help the United Nations deliver food and water to people suffering from droughts?

A map! Not just any map, but an online map with accurate information. In **crisis mapping,** Patrick Meier combines information from government and international agencies with tweets and text messages sent by volunteers. He and his team use this information to update an online map of the areas where disasters have struck. Crisis mapping saves lives by providing rescue workers with an up-to-the-minute picture of what is going on and where help is needed.

After the 2010 earthquake in Haiti, Patrick's efforts assisted citizens, aid workers, and the U.S. Coast Guard. Patrick's team of volunteers mapped the impact of the earthquake in near real time, providing professionals with the most up-to-date information available.

Patrick continues to help governments and aid groups respond to natural disasters. He has used his crisis mapping skills around the world and wrote a book based on them. Patrick is also researching ways that robotic equipment can be used to help after a disaster.

DCI ESS3.B: Natural Hazards. A variety of hazards result from natural processes (e.g., earthquakes, tsunamis, volcanic eruptions). Humans cannot eliminate the hazards but can take steps to reduce their impacts. (4-ESS3-2)
CETS Influence of Engineering, Technology, and Science on Society and the Natural World. Engineers improve existing technologies or develop new ones to increase their benefits, to decrease known risks, and to meet societal demands. (4-ESS3-2)

Explorer

Patrick Meier is a leader in the field of crisis mapping. Patrick didn't always think he would be working with new types of maps. In school he liked the subjects computer science and philosophy. Later in college he studied humanitarian affairs. In his work he combines his love of geography and technology in a way that helps people around the world.

NGL Science How did you get into mapping?

Patrick Meier When I was 12, the first Gulf War broke out. I had a big map of the Middle East and started physically mapping the updates with crayons and pens and markers.

NGL Science What are some events you and your team have tracked?

Patrick Meier Haiti started it all. A month later there was an earthquake in Chile. Then the floods in Pakistan that summer. Russian fires in July. Floods in Brisbane in January. A major earthquake in Christchurch, New Zealand, that February.

NGL Science What are people saying about your crisis-mapping technology?

Patrick Meier Many humanitarian organizations say our crisis-mapping technology is revolutionizing disaster relief efforts.... Now we can pinpoint urgent needs instantly, saving time and lives.

Check In 📓 My Science Notebook

Congratulations! You have completed *Earth Science*. Let's reflect on what you have learned. Here is a checklist to help you judge your progress. Look through your science notebook to find examples for each statement in the checklist. What could you do better? Write it on a separate page in your science notebook.

▼ Read each item in this list. Ask yourself if you think you did a good job of it.

For each item, select the choice that is true for you: A. Yes B. Not Yet

- I defined and illustrated science vocabulary, science concepts, and main ideas.
- I labeled drawings. I included captions and notes to explain ideas.
- I collected objects such as photos and magazine or newspaper clippings.
- I used tables, charts, or graphs to record observations and data in investigations.
- I recorded evidence for explanations and conclusions in investigations.
- I described how scientists and engineers answer questions and solve problems.
- I asked new questions.
- I did something else. (Tell about it.)

Reflect on Your Learning 📓 My Science Notebook

1. How can you use patterns to help you understand one main science idea?
2. Choose one main science idea that you think was most important to learn about. Explain your thinking.

More to Explore

I became a biomedical engineer because I wanted to help people. I noticed a problem and wanted to solve it. As a scientist, I have learned to make observations, do investigations, study evidence, and design solutions. I have made careful measurements and tests. I have recorded my results in my notebook. I have shared my results with others.

Write about how you used your science notebook. Tell how it helped you ask questions and find answers about the natural world. What were some of the most surprising things you learned? How did new evidence help you form new explanations? Discuss what you learned with your classmates. And remember to keep on exploring. Never be afraid to ask new questions. Always be curious!

David works with these young scientists as part of a program he runs in Sierra Leone. The program gets young people involved in solving local problems in that country.

Science Safety

Be responsible, look, and listen in the science lab. Follow all lab safety rules to stay safe. Know what procedure to follow for each lab you conduct. If anything is unclear, ask an adult for help. Always be aware of your space. Report an accident to an adult immediately. Tell your teacher about any allergies you have.

The Lab Space
- Know where the first aid kit is kept.
- Know the location of the fire blanket.
- Sit only on lab chairs or stools, never lab tables.
- Do not run.
- Keep your area neat.

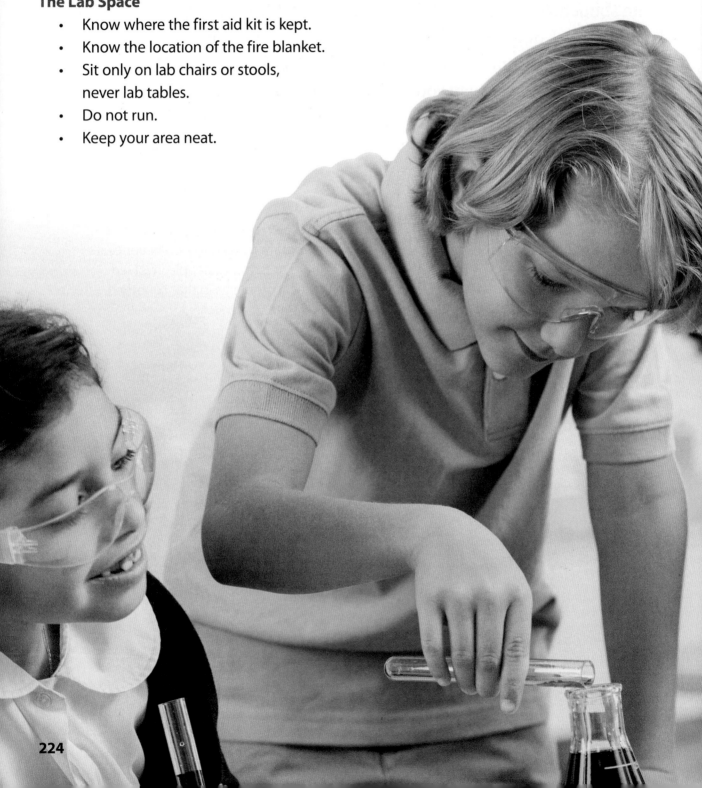

Lab Clothing
- Tie back loose or long hair.
- Do not let jewelry or clothing hang loosely.
- Shoes should cover the foot; no sandals.
- Wear goggles, gloves, or a lab apron when told.

Animal and Plant Safety
- Be aware of the living things in your lab.
- Wash hands after handling plant or animal material.

Chemical Safety
- Do not eat or drink anything in the science room.
- Never mix chemicals.
- Find out where to dispose of chemicals.
- Keep hands away from your eyes and mouth.
- Wash your hands when finished.

Fire and Electrical Safety
- Do not touch charged ends of batteries.
- Never connect multiple batteries in a circuit.

Glass Safety
- Ask an adult to place broken glass in a sealed container.

Cleanup
- Close all containers.
- Return materials to their correct storage locations.
- Throw out used gloves.
- Wash hands with soap and water.

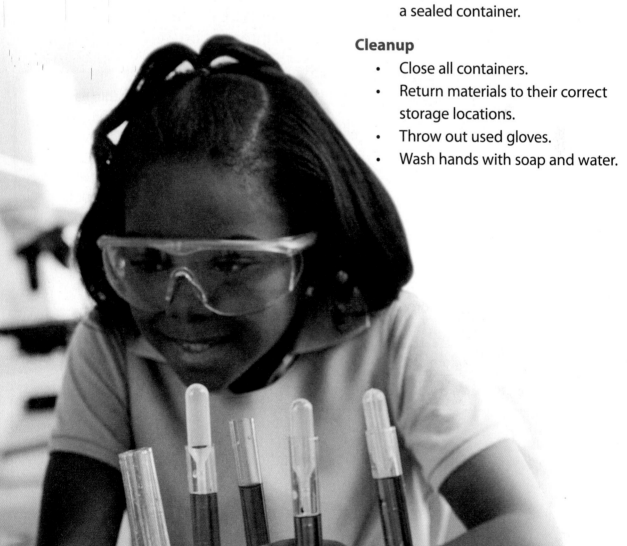

Tables and Graphs

When scientists and engineers make observations, they often record different kinds of information. This is called data. Data can be measurements or facts that tell about observations.

Data help scientists and engineers answer questions, make predictions, and ask new questions. Scientists and engineers need to organize their data to make sense of the data and share with others. Tables and graphs are tools for organizing, summarizing, and sharing data. You can use these tools to organize data, too.

Tables

A table is a set of **rows** and **columns.** Rows and columns set up a simple grid that makes a place for every piece of information.

PRECIPITATION IN WESTERN U.S. CITIES

City, State	Precipitation (cm (in.))
Las Vegas, Nevada	10 cm (4 in.)
Phoenix, Arizona	20 cm (8 in.)
Riverside, California	25 cm (10 in.)
San Diego, California	25 cm (10 in.)
Los Angeles, California	30 cm (12 in.)
Denver, Colorado	38 cm (15 in.)
San Jose, California	40 cm (16 in.)
Salt Lake City, Utah	40 cm (16 in.)

Source: National Weather Service, 2016

This table has two columns. One column shows cities and states. The other column gives amounts of precipitation.

Cite the source of your data, if you did not collect it yourself. This shows others how to confirm whether your data are valid, or reliable.

A table is a useful way to organize a list of information. You can add new information to a table by adding more rows or columns. What information does this table help organize? How is it organized?

Bar Graphs

A bar graph is useful for comparing different amounts of similar information.

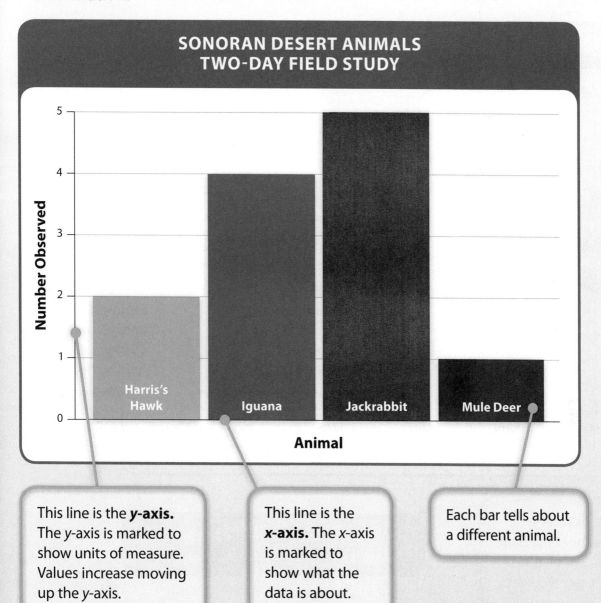

This line is the **y-axis.** The y-axis is marked to show units of measure. Values increase moving up the y-axis.

This line is the **x-axis.** The x-axis is marked to show what the data is about.

Each bar tells about a different animal.

Each bar is a different height. Bar height answers the question "How many?" The top of each bar lines up with a unit of measure on the y-axis. Which animal species was observed the most on a visit to the Sonoran Desert?

A pictograph uses a series of pictures instead of bars. What pictures might you use to make a pictograph of this information?

Line Graphs

A line graph is good for showing how data changes over time. Like a bar graph, a line graph has an *x*-axis and a *y*-axis.

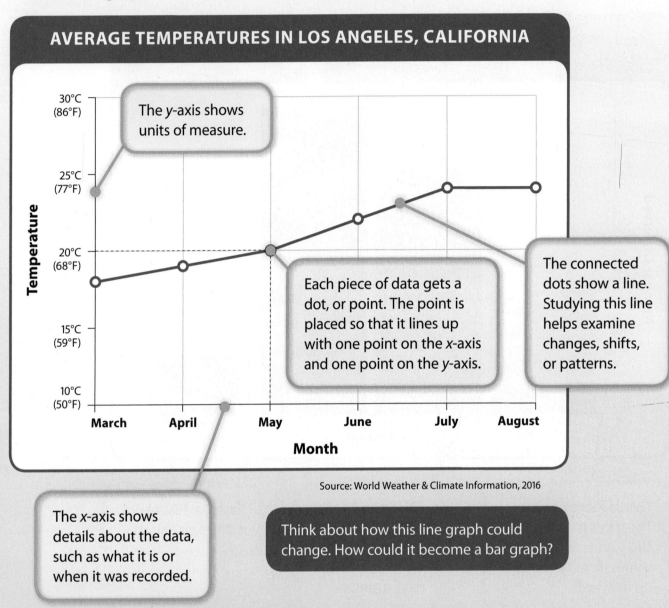

Source: World Weather & Climate Information, 2016

Think about how this line graph could change. How could it become a bar graph?

A line graph might look at data across time, across ages, or across distances. This line graph shows how temperature changes across several months. In what month does the coolest temperature occur? What pattern do you see across the months?

Circle Graphs

Circle graphs show data as parts that make up a whole. These graphs are also called "pie graphs" because the parts look like pie slices. The whole circle stands for the total amount, or 100%, of the data collected. A circle graph is useful for data that tell how much or how many of something.

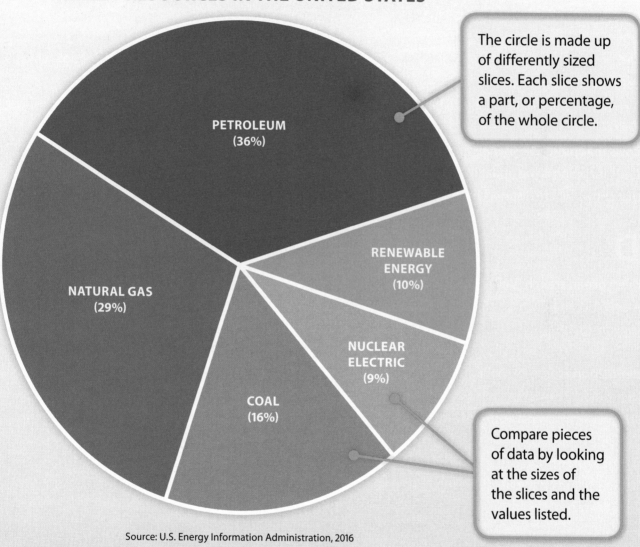

ENERGY RESOURCES IN THE UNITED STATES

- PETROLEUM (36%)
- NATURAL GAS (29%)
- COAL (16%)
- RENEWABLE ENERGY (10%)
- NUCLEAR ELECTRIC (9%)

The circle is made up of differently sized slices. Each slice shows a part, or percentage, of the whole circle.

Compare pieces of data by looking at the sizes of the slices and the values listed.

Source: U.S. Energy Information Administration, 2016

The blue slices in this graph show how much of the total use came from nonrenewable energy resources. The green slice shows how much came from renewable energy resources. Which energy resource is used the most? Which energy resource is used the least? How much of the energy used came from nonrenewable resources?

A

amplitude (AMP-li-tüd)
Amplitude is the distance between the crest or trough and the middle point of a wave. (p. 78)

C

column (KO-lum)
A column is a vertical section of a table. (p. 226)

crisis mapping (CRĪ-sis MAP-ing)
Crisis mapping is the updating of an online map that shows where disasters have struck. (p. 220)

D

deep ocean trench (DĒP Ō-shun TRENCH)
A deep ocean trench is a steep underwater canyon. (p. 199)

deposition (de-pe-ZI-shun)
Deposition is the laying down of sediment and rock in a new place. (p. 161)

digitize (DIJ-i-tīz)
To digitize means to put information in digital code form. (p. 86)

E

earthquake (URTH-kwāk)
An earthquake is the shaking of the ground caused by the movement of Earth's crust. (p. 178)

electric circuit (i-LEK-trik SUR-kit)
An electric circuit is a complete path through which an electric current can pass. (p. 46)

electric current (i-LEK-trik KUR-ent)
Electric current is the transfer of electrical energy through a material. (p. 44)

electrical energy (i-LEK-trik-ul EN-er-jē)
Electrical energy is the energy of charged particles. (p. 44)

energy (EN-er-jē)
Energy is the ability to do work. (p. 22)

energy of motion (EN-er-jē uv MŌ-shun)
Energy of motion is the energy that is present when an object moves. (p. 50)

Gravity pulls Earth and the skydiver toward each other.

erosion (i-rō-zhun)
Erosion is the moving of sediment from one place to another. (p. 160)

erupt (ē-RUPT)
To erupt is to release melted rock, ash, and gases up through a volcano onto Earth's surface. (p. 186)

evacuate (ē-VAK-yū-āt)
To evacuate means to move from a place of danger to a safe place. (p. 190)

evidence (EV-i-dens)
A piece of evidence is an observation that supports an idea or conclusion. (p. 10)

experiment (eks-PAIR-i-ment)
In an experiment, you change only one variable, measure or observe another variable, and control other variables so they stay the same. (p. 12)

F

fault (FAWLT)
A fault is a break in Earth's surface where huge slabs of rock come together. (p. 180)

fossil (FO-sil)
A fossil is a trace of a plant or animal that lived long ago. (p. 210)

fossil fuel (FO-sil FYŪL)
A fossil fuel is a source of energy that formed from the remains of plants and animals that lived millions of years ago. (p. 68)

G

glacier (GLĀ-shur)
A glacier is a huge area of slow-moving ice. (p. 168)

global positioning system (GPS)
(GLŌ-bul pe-ZI-shun-ing SIS-tem)
A global positioning system (GPS) is a tool that uses satellites to locate position. (p. 86)

gravity (GRA-vi-tē)
Gravity is a force that pulls things toward the center of Earth. (p. 172)

H

hazard (HA-zurd)
A hazard is something that is harmful or dangerous. (p. 178)

hypothesis (hi-POTH-uh-sis)
A hypothesis is a statement giving a possible answer to a question that can be tested by an experiment. (p. 12)

I

infer (in-FUR)
When you infer, you use what you know and what you observe to draw a conclusion. (p. 12)

investigate (in-VES-ti-gāt)
You investigate when you carry out a plan to answer a question. (p. 12)

L

landslide (LAND-slīd)
A landslide is a rapid movement of rock, soil, and other material down a hill or mountain. (p. 173)

lava (LAH-vah)
Lava is melted rock that flows from a volcano onto Earth's surface. (p. 186)

liquefaction (li-qui-FAK-shun)
Liquefaction is the change of a solid area of ground into a liquid-like surface. (p. 182)

longitudinal wave (long-ji-TOOD-u-nel WĀV)
In a longitudinal wave, particles of material vibrate back and forth in the same direction as the direction of the wave. (p. 80)

M

magma (MAG-mah)
Magma is melted rock inside Earth. (p. 186)

mid-ocean ridge (MID-ō-shun RIDJ)
A mid-ocean ridge is an underwater mountain chain made up of thousands of volcanic peaks. (p. 199)

model (MO-del)
In science, models are used to explain or predict natural events. A model can show how a process works in real life. (p. 12)

motion (MŌ-shun)
When an object is moving, it is in motion. (p. 26)

N

nonrenewable energy resource (non-rē-NŪ-i-bel EN-er-jē rē-SORS)
A nonrenewable energy resource is an energy resource that will eventually run out. (p. 68)

O

observe (ub-ZURV)
When you observe, you use your senses to gather information about an object or event. (p. 10)

organism (OR-ga-niz-um)
An organism is a living thing. (p. 170)

P

pistil (PIS-til)
A pistil is the part of the flower that forms fruit. (p. 113)

pitch (PICH)
The pitch of a sound is how high or low the sound is. (p. 80)

R

reflect (rē-FLEKT)
To reflect is to bounce light off of an object. (p. 128)

renewable energy resource (rē-NŪ-i-bel EN-er-jē rē-SORS)
A renewable energy resource is an energy resource that will never run out. (p. 70)

rift (RIFT)
A rift is an opening where pieces of Earth's crust are being pulled apart. (212)

row (RŌ)
A row is a horizontal section of a table. (p. 226)

The **pistil** and **stamen** are parts of the flower that help the plant reproduce.

S

sand dune (SAND DŪN)
A sand dune is a landform caused when wind deposits sand. (p. 162)

sediment (SE-di-ment)
Sediment is material that comes from the weathering of rock. (p. 158)

sedimentary rock (se-di-MENT-u-rē ROK)
Sedimentary rock is rock formed from sediment from the weathering and erosion of rock on Earth's surface. (p. 210)

seismograph (SĪZ-mō-graf)
A seismograph is a device that records earthquake activity. (p. 190)

seismometer (sīz-MO-mi-ter)
A seismometer is a tool that measures earthquake activity. (p. 190)

solar energy (SOL-er EN-er-jē)
Solar energy is heat and light energy from the sun. (p. 70)

stamen (STĀ-min)
A stamen is a plant part that makes pollen. (p. 113)

T

thermal energy (THUR-mel EN-er-jē)
Thermal energy is the energy of heat. (p. 40)

transfer (TRANS-fer)
To transfer is to pass from one object to another. (p. 26)

transform (TRANS-form)
To transform is to change. (p. 30)

transmit (trans-MIT)
To transmit means to send. (pp. 86)

transverse wave (tranz-vers WĀV)
In a transverse wave, particles of material move up and down as the wave moves side to side. (p. 76)

tsunami (soo-NAH-mē)
A tsunami is a series of ocean waves caused by an underwater earthquake, volcano, or landslide. (p. 179)

V

variable (VAIR-ē-u-bl)
A variable is a factor that can change or be controlled in an experiment, investigation, or model. (p. 12)

vibration (vī BRĀ shun)
A vibration is a rapid, back-and-forth movement. (p. 30)

volcano (vol-CĀ-nō)
A volcano is formed when molten rock deep inside Earth rises to the surface. (p. 178)

volume (VOL-yūm)
The volume of sound is the measure of loudness. (p. 80)

W

wave (WĀV)
A wave transfers energy from one place or object to another. (p. 76)

wavelength (WĀV-length)
A wavelength is the distance from one crest to the next crest in a wave. (p. 78)

weathering (WE-thur-ing)
Weathering happens when rocks break apart, wear away, or dissolve into smaller particles, or when the materials in the rock are changed. (p. 158)

wind energy (WIND EN-er-jē)
Wind energy is energy produced from the motion of wind. (p. 70)

X

x-axis (EKS AKSIS)
The x-axis on a graph is the horizontal base line. The x-axis is usually presented on the bottom of the graph. (p. 227)

Y

y-axis (WĪ AKSIS)
The y-axis on a graph is the vertical base line. The y-axis is usually presented on the left side of the graph. (p. 227)

Weathering of rock can produce unusual shapes.

Program Consultants

Randy L. Bell, Ph.D.
Associate Dean and Professor of Science Education, College of Education, Oregon State University

Malcolm B. Butler, Ph.D.
Professor of Science Education and Associate Director; School of Teaching, Learning and Leadership; University of Central Florida

Kathy Cabe Trundle, Ph.D.
Department Head and Professor, STEM Education, North Carolina State University

Judith S. Lederman, Ph.D.
Associate Professor and Director of Teacher Education, Illinois Institute of Technology

Center for the Advancement of Science in Space, Inc.
Melbourne, Florida

Acknowledgments

Grateful acknowledgment is given to the authors, artists, photographers, museums, publishers, and agents for permission to reprint copyrighted material. Every effort has been made to secure the appropriate permission. If any omissions have been made or if corrections are required, please contact the Publisher.

 is a registered trademark of Achieve. Neither Achieve nor the lead states and partners that developed the Next Generation Science Standards was involved in the production of, and does not endorse, this product.

Photographic and Illustrator Credits
Front cover wrap ©Alberto Ghizzi Panizza/500px
Back cover Lynn Johnson/National Geographic Creative

Acknowledgments and credits continue on page 244.

Copyright © 2019 Cengage Learning, Inc.

ALL RIGHTS RESERVED. No part of this work covered by the copyright herein may be reproduced or distributed in any form or by any means, except as permitted by U.S. copyright law, without the prior written permission of the copyright owner.

"National Geographic", "National Geographic Society" and the Yellow Border Design are registered trademarks of the National Geographic Society® Marcas Registradas

> For product information and technology assistance, contact us at
> Customer & Sales Support, 888-915-3276
> For permission to use material from this text or product, submit all requests online at **www.cengage.com/permissions**
> Further permissions questions can be emailed to
> **permissionrequest@cengage.com**

National Geographic Learning | Cengage
1 N. State Street, Suite 900
Chicago, IL 60602

National Geographic Learning, a Cengage company, is a provider of quality core and supplemental educational materials for the PreK-12, adult education, and ELT markets. Cengage is a leading provider of customized learning solutions with employees residing in nearly 40 different countries and sales in more than 125 countries around the world. Find your local representative at **NGL.Cengage.com/RepFinder**.

Visit National Geographic Learning online at **NGL.Cengage.com/school**

ISBN: 978-13379-10255

Printed in the United States of America
Print Number: 04
Print Year: 2022